The Anti-Gravity Handbook

The Lost Science Series:
The Anti-Gravity Handbook
Anti-Gravity & the World Grid
Anti-Gravity & the Unified Field
The Free-Energy Device Handbook
The Energy Grid by Bruce Cathie
The Bridge To Infinity by Bruce Cathie
The Harmonic Conquest of Space by Bruce Cathie
Vimana Aircraft of Ancient India & Atlantis
Eher Technology by Rho Sigma
The Fantastic Inventions of Nikola Tesla
Man-Made UFOs: 1944-1994
UFOs and Anti-Gravity: Piece For A Jig-Saw
Flying Saucers Over Los Angeles

The Lost Cities Series:
Lost Cities of Atlantis, Ancient Europe & the Mediterranean
Lost Cities of North & Central America
Lost Cities & Ancient Mysteries of South America
Lost Cities of Ancient Lemuria & the Pacific
Lost Cities & Ancient Mysteries of Africa & Arabia
Lost Cities of China, Central Asia & India

The Mystic Traveller Series:
In Secret Mongolia by Henning Haslund (1934)
Men & Gods In Mongolia by Henning Haslund (1935)
In Secret Tibet by Theodore Illion (1937)
Darkness Over Tibet by Theodore Illion (1938)
Atlantis In Spain by E.M. Whishaw (1928)
Mystery Cities of the Maya by Thomas Gann (1925)
In Quest of Lost Worlds by Byron de Prorok (1937)

Write for our free catalog of unusual books and videos.

THE ANTI-GRAVITY HANDBOOK

EDITED BY
DAVID HATCHER CHILDRESS

Adventures Unlimited Press

This book is dedicated to the scientists
and engineers who continue to forge ahead
despite opposition from all sides.

The Anti-Gravity Handbook
Revised Edition

Copyright 1998
Adventures Unlimited Press

First Edition ©1985

Printed in the United States of America

ISBN 0-932813-20-8

Published by
Adventures Unlimited Press
One Adventure Place
Kempton, ILLINOIS 60946 USA

Table of Contents

1. Arthur C. Clarke On Anti-Gravity — 3
   by D. Hatcher Childress
2. How To Build A Flying Saucer — 7
   by T. B. Pawlicki
3. The Anti-Gravity Equation — 25
   by Captain Bruce Cathie
4. Profiles of Anti-Gravity — 37
   Researchers #1: Albert Einstein
5. Up In The Air Over Anti-Gravity — 67
   by W. P. Donavan
6. Electro-magnetic Pulse — 69
   by Karl Kruszelenski
7. Moller Discoid Aircraft — 73
8. A Machine to End War — 81
   by Nikola Tesla
9. Gravity Nullified-Crystal Resonances & Anti-Gravity — 87
10. Eleven Things You Never Knew About The Moon — 97
11. NASA, the Moon and Anti-Gravity — 101
    by D. Hatcher Childress
12. Mars, NASA, and Anti-Gravity — 109
    by D. Hatcher Childress
13. Dr. Zitzenpop's Generic Anti-Gravity Equation — 121
    by W. P. Donavan
14. Anti-Gravity Fashions of the 30's — 125
15. Ancient Indian Aircraft Technology — 129
    D. Hatcher Childress
16. A Selection of Anti-Gravity Patents — 137
17. Zeppelins, UFOs, Anti-Gravity and the Mysteries of the Airship Age — 147
    by D. Hatcher Childress
18. Newpaper Headlines of the Past, Present and Future Or "Aliens Stole My Baby" — 155
19. Levitation: Personal Defiance of Gravity — 167
    by D. Hatcher Childress
20. Anti-Gravity Comix and Classified Ads — 173
21. Meanwhile: Back In The Future — 185
    by W. P. Donavan

Appendix
   The Worlds' Most Complete Bibliography
   on Gravity Control and UFO Material

THANKS to Bill, Bert, Jean-Louis, Marge, Paul, Larry, Molly, Jake, Mom, Dad, Alan, Tim, Stephanie, Warren, Jamie, Frank and galaxy of other folks, who somehow helped create this book, whether they know it or not.

Someday, after we have mastered the winds, the waves, the tides and gravity, we shall harness for God the energies of love. Then for the second time in the history of the world man will have discovered fire.

   -Teilhard de Chardin

The Anti-Gravity Handbook

"Authorities", "disciples", and "schools" are the curse of science and do more to interfere with the work of the scientific spirit than all its enemies.

-T. H. Huxley

**There isn't any energy crisis.
It's simply a crisis of ignorance.
   -R. Buckminster Fuller**

ARTHUR C. CLARKE ON ANTI-GRAVITY

Profile by David H. Childress

"Of all natural forces, gravity is the most mysterious and the most implacable. It controls our lives from birth to death, killing or maiming us if we make the slightest slip. No wonder that, conscious of their earthbound slavery, men have always looked wistfully at birds and clouds, and have pictured the sky as the abode of the gods. The very expression 'heavenly being' implies a freedom from gravity which, at the present, we have known only in our dreams."

So wrote Arthur C. Clarke in the 1963 book, "Profiles of the Future." One of the most famous science fiction authors of all time, Clarke is also the author of a number of books on science, including such classics as "Interplanetary Flight," "The Exploration of Space," "Voices from the Sky" and "Profiles of the Future." Probably his best known work is "2001: A Space Odyssey."

Clarke, who was left totally paralyzed in 1962 as the result of a spinal injury, has recovered most of his movement, and now lives in Sri Lanka. Clarke has always been interested in space, and in fact said in a recent interview, "Space brought me to Sri Lanka. I was interested in diving here simply because it is the only way of reproducing the condition of weightlessness, which is characteristic of space flight."

Clarke became something of a visionary when he predicted a manned lunar landing by 1970 in the early sixties, a prediction which came true. Clarke predicted "Gravity Control" by the year 2050, contact with extraterrestrials by 2030, the colonization of the planets by 2000 and the discovery of gravity waves and interplanetary landings by 1980.

Yet Clarke is skeptical of UFOs and ESP. He finds most demonstrators of clairvoyance ESP and the like to be charlatans who should be exposed. Clarke sees no evidence for UFOs and once stated, "We should only be concerned with close encounters. Either they exist or they don't. If anyone reported a Tyrannosaurus Rex loose in Central Park I'd verify it quite quickly; the same goes for flying saucers." Clarke would seem to find all governments benevolent, and cover-ups as much a hoax as ESP and UFOs.

Clarke does, however, believe that anti-gravity is a possibility. "When the history of the human race is written... perhaps our space faring descendants will be as little concerned with gravity as were our remote ancestors, when they floated effortlessly in the buoyant sea."

The problem of gravitation concerns Clarke quite a bit. It is the one force that we cannot duplicate, nor even sufficiently understand. He states that as yet, we cannot generate gravity at all, though he is apparently unaware of the Searle Disk and the "Biefield/Brown Effect," two experiments that predate his article on gravity control.

The extreme weakness of gravity makes it even more exasperating that we cannot control it, says Clarke. In a statement that might seem gross arrogance, Clarke states "No competent scientist, at this state of our ignorance, would deliberately set out to look for a way of overcoming gravity." Is Clarke totally unaware of the work of Nikola Tesla, T. Townsend Brown, John Searle, The Philadelphia Experiment, and even Albert Einstein? Is todays science too inept to tackle one of the basic foundations of physics?

Clarke does at least agree with those physicists and mathematicians that are at least trying to research the subject and gain a basic knowledge of what gravity is, something we know precious little about, at least officially. Clarke seems to agree with the statement of Dr. John Pierce at the Bell Tel- ephone Laboratories, "Anti-Gravity is for the birds." Yet Clarke believes we will conquer gravity, but that we do not have the knowledge yet.

Yet, possibly, Clarke is just being conservative so as not to arouse the wrath of the extremely conservative scientists who make up the bulk of academia. Also, Clarke was speaking more than twenty years ago when he and many scientists had much to discover.

Clarke, at one point in "Profiles of the Future" says that "it seems most unlikely that there exist gravitational fields, anywhere in the universe, more than a few hundred thousand times more powerful than earth's." A few years later neutron stars and black holes were discovered, which totally blew that hypothesis out of the water.

Clarke relates an interesting survey done in 1960 by the Harvard Business Review called a "Survey on the Space Program," which received almost two thousand replies to it's detailed five-page questionnaire to businessmen and company executives.

When asked to relate the degree of probability of various by-products of space research, the executives' votes for anti-gravity came out: almost certain, 11 percent; very likely, 21 percent; never will happen, 6 percent. In fact, they related anti-gravity as rather more likely than mining or colonizing the planets!

Clarke goes on to say in "Profiles of Science" that those tinkerers with anti-gravity are dreaming. Yet, for someone who doesn't understand gravity himself (no one understands gravity according to Clarke), it seems absurd that he should ridicule someone who may have a different conception of gravity than himself.

Yet Clarke is not so stuffy as he seems. After conditioning the reader to gravity control being a long way off, he tells of the fantastic inventions and uses for anti-gravity once we have invented it, which we will one day, he assures us. "Despite the above-mentioned skepticism of the physicists there seems no fundamental impossibility about such a device."

Clarke gets lively on his discussions of gravity screens and negative-matter. He conjures a vision of a space prospector in his space jeep gathering negative-matter from an asteroid (which of course, would emit a negative-gravity field) and having a hell of a time trying to get back to earth due to the earth's gravity repelling his cargo. He concludes that negative-gravity substances, should they exist, would have a limited use.

He also sees the possibility of degravitizing ordinary substances, and finally concludes that any form of gravity would also be a propulsion system. "We should expect this, as gravity and acceleration are so intimately linked."

"If anti-gravity devices turn out to be bulky and expensive, their use will be limited to fixed installations and to large vehicles--perhaps the size of which we have not yet seen on this planet," says Clarke.

"One can imagine the bulk movement of freight or raw materials along 'gravity pipelines', directed and focused fields in which objects would be supported and would move like iron toward a magnet. Our descendants may be quite accustomed to seeing their goods and chattels sailing from place to place without visible means of support."

Clarke even sees the development of the portable gravity-control unit, so compact that a man could strap it on his shoulders or around his waist. He sees it as even being a permanent part of his clothing, taken as much for granted as wristwatch or personal transceiver. It could be used to reduce his apparent weight down to zero, or to provide propulsion.

Thus, the one-man gravitator would be perhaps the most revolutionary invention of all time, he thinks. One would not need an elevator in a city if there were a convenient window. Clarke sees these "levitators" as doing for mountains what aqualungs have done for the sea. Tourists will be floating over the Himalayas, and even the summit of Mount Everest may be a popular tourist spot.

Arthur C. Clarke may be a true visionary, and his only problem may be that he is somewhat myopic. His dreams of conquering gravity may be much closer than he realizes; in fact they may have already come true!

T. B. PAWLICKI

HOW TO BUILD A FLYING SAUCER

After So Many Amateurs Have Failed

At the end of the nineteenth century, the most distinguished scientists and engineers declared that no known combination of materials and locomotion could be assembled into a practical flying machine. Fifty years later another generation of distinguished scientists and engineers declared that it was technologically infeasible for a rocket ship to reach the moon. Nevertheless, men were getting off the ground and out into space even while these words were uttered.

In the last half of the twentieth century, when technology is advancing faster than reports can reach the public, it is fashionable to hold the pronouncements of yesteryear's experts to ridicule. But there is something anomalous about the consistency with which eminent authorities fail to recognize technological advances even while they are being made. You must bear in mind that these men are not given to making public pronouncements in haste; their conclusions are reached after exhaustive calculations and proofs, and they are better informed about their subject than anyone else alive. But by and large, revolutionary advances in technology do not contribute to the advantage of established experts, so they tend to believe that the challenge cannot possibly be realized.

The UFO phenomenon is a perversity in the annals of revolutionary engineering. On the one hand, public authorities deny the existence of flying saucers and prove their existence to be impossible. This is just as we should expect from established experts. But on the other hand, people who *believe* that flying saucers exist have produced findings that only tend to prove UFOs are technologically infeasible by any known combination of materials and means of locomotion.

There is reason to suspect that the people who believe in the existence of UFOs do not want to discover the technology because it is not in the true believer's self-interest that a flying saucer be within the capability of human engineering. The true believer wants to believe that UFOs are of extraterrestrial origin because he is seeking some kind of relief from debt and taxes by an alliance with superhuman powers.

If anyone with mechanical ability really wanted to know how a saucer flies, he would study the testimonies to learn the flight characteristics of the craft, and then ask, "How can we do this saucer thing?" This is probably what Wernher Von Braun said when he decided that it was in his self-interest to launch man into space: "How can we get this bird off the ground, and keep it off?"

Well, what is a flying saucer? It is a disc-shaped craft about thirty feet in diameter with a dome in the center accommodating the crew and, presumably, the operating machinery. And it flies. So let us begin by building a disc-shaped air foil, mount the cockpit and the engine under a central canopy, and see if we can make it fly. As a matter of fact, during World War II the United States actually constructed a number of experimental aircraft conforming to these specifications, and photographs of the craft are published from time to time in popular magazines about science and flight. It is highly likely that some of the UFO reports before 1950 were sightings of these test flights. See how easy it is when you *want* to find answers to a mystery?

The mythical saucer also flies at incredible speeds. Well, the speeds believed possible depend upon the time and the place of the observer. As stated earlier, a hundred years ago, twenty-five miles per hour was legally prohibited in the belief that such terrific velocity would endanger human life. So replace the propellor of the experimental disc airfoil with a modern aerojet engine. Is Mach 3 fast enough for believers?

But the true saucer not only flies, it also hovers. You mean like a Hovercraft? One professional engineer translated Ezekiel's description of heavenly ships as a helicopter-cum-Hovercraft.

But what about the anomalous electromagnetic effects manifest in the space surrounding a flying saucer? Well, Nikola Tesla demonstrated a prototype of an electronic device that was eventually developed into the electron microscope, the television screen, and an aerospace engine called the Ion Drive. Since World War II, the engineering of the Ion Drive has been advanced as the most promising solution to the propulsion of interplanetary spaceships. The Drive operates by charging atomic particles and directing them with electromagnetic force as a jet to the rear, generating a forward thrust in reaction. The advantage of the Ion Drive over chemical rockets is that a spaceship can sweep in the ions it needs from its flight path, like an aerojet sucks in air through its engines. Therefore, the ship must carry only the fuel it needs to generate the power for its chargers; there is no need to carry dead weight in the form of rocket exhaust. There is another advantage to be derived from ion rocketry: The top speed of a reaction engine is limited by the ejection velocity of its exhaust. An ion jet is close to the speed of light. If space travel is ever to be practical, transport will have to achieve a large fraction of the speed of light.

In 1972 the French journal *Science et Avenir* reported Franco-American research into a method of ionizing the airstream flowing over wings to eliminate the sonic boom, a serious objection to the commercial success of the Concorde. Four years later a picture appeared in an American tabloid of a model aircraft representing the state of current development. The photograph shows a disc-shaped craft, but not so thin as a saucer; it looks more like a flying curling stone. In silent flight, the ionized air flowing around the craft glows as a proper UFO should. The last word comes from an engineering professor at the local university; he has begun the construction of an Ion-Drive Flying Saucer in his backyard.

To the true believer, the flying saucer has no jet. It seems to fly by some kind of antigravity. As antigravity is not known to exist in physical theory or experimental fact in popular science, the saucer is clearly alien and beyond human comprehension. But *antigravity* depends upon what you conceive *gravity* to be, doesn't it?

For all practical purposes, you do not have to understand what Newton and Einstein mean by gravity. Gravity is an acceleration downward, to the center of the earth. Therefore, antigravity is an acceleration upward. As far as practical engineering is concerned, any means to achieve a gain in altitude is an antigravity engine. An airplane is an antigravity engine, a balloon is an antigravity engine, a rocket is an antigravity engine, a stepladder is an antigravity engine. See how easy it is to invent an antigravity engine?

There are three basic kinds of locomotive engines. The primary principle is traction. The foot and the wheel are traction engines. The traction engine depends upon friction against a surrounding medium to generate movement, and locomotion can proceed only as far and as speedily as the surrounding friction will provide. The secondary principle is displacement. The balloon and the submarine rise by displacing a denser medium; they descend by displacing less than their weight. The tertiary drive is the rocket engine. A rocket is driven by reaction from the mass of material it ejects. Although a rocket is most efficient when not impeded by a surrounding medium, it must carry not only its fuel but also the mass it must eject. As a consequence, the rocket is impractical where powerful acceleration is required for extended drives. In chemical rocketry, ten minutes is a long burn for powered flight. What is needed for practical antigravity locomotion is a fourth principle which does not depend upon a surrounding medium or ejection of mass.

You must take notice that none of the principles of locomotion required any new discovery. They have all been around for thousands of years, and engineering only implemented the principle with increasing efficiency. A fourth principle of locomotion has also been around for thousands of years: It is centrifugal force. Centrifugal force is the principle of the military sling and medieval catapult.

Everyone knows that centrifugal force can overcome gravity. If directed upward, centrifugal force can be used to drive an antigravity engine. The problem engineers have been unable to solve is that centrifugal force is generated in all directions on the plane of the centrifuge. It won't provide locomotion unless the force can be concentrated in one direction. The solution of the sling, of releasing the wheeling at the instant the centrifugal force is directed along the ballistic trajectory, has all the inefficiencies of a cannon. The difficulty of the problem is not real, however. There is a mental block preventing people from perceiving a centrifuge to be anything other than a flywheel.

A bicycle wheel is a flywheel. If you remove the rim and tire, leaving only the spokes sticking out from the hub, you still have a flywheel. In fact, spokes alone make a more efficient flywheel than the complete wheel; this is because momentum goes up only in proportion to mass but with the square of speed. Spokes are made of drawn steel with extreme tensile strength, so spokes alone can generate the highest levels of centrifugal force long after the rim and tire have disintegrated. But spokes alone still generate centrifugal force equally in all directions from the plane of rotation. All you have to do to concentrate centrifugal force in one direction is remove all the spokes but one. That one spoke still functions as a flywheel, even though it is not a wheel any longer.

See how easy it is once you accept an attitude of solving one problem at a time as you come to it? You can even add a weight to the end of the spoke to increase the centrifugal force.

But our centrifuge still generates a centrifugal acceleration in all directions around the plane of rotation even though it doesn't generate acceleration equally in all directions at the same time. All we have managed to do is make the whole ball of wire wobble around the common center of mass between the axle and the free end of the spoke. To solve this problem, now that we have come to it, we need merely to accelerate the spoke through a few degrees of arc and then let it complete the cycle of revolution without power. As long as it is accelerated during the same arc at each cycle, the locomotive will lurch in one direction, albeit intermittently. But don't forget that the piston engine also drives intermittently. The regular centrifugal pulses can be evened out by mounting multiple centrifuges on the same axle so that a pulse from another flywheel takes over as soon as one pulse of power is past its arc.

The next problem facing us is that the momentum imparted to the centrifugal spoke carries it all around the cycle with little loss of velocity. The amount of concentrated centrifugal force carrying the engine in the desired direction is too low to be practical. Momentum is half the product of mass multiplied by velocity squared. Therefore, what we need is a spoke that has a tremendous velocity with minimal mass. They don't make spokes like that for bicycle wheels. A search through the engineers' catalog, however, turns up just the kind of centrifuge we need. An electron has no mass at rest (you cannot find a smaller minimum mass than that); all its mass is inherent in its velocity. So we build an electron raceway in the shape of a doughnut in which we can accelerate an electron to a speed close to that of light. As the speed of light is approached, the energy of acceleration is converted to a momentum approaching infinity. As it happens, an electron accelerator answering our need was developed by the University of California during the last years of World War II. It is called a betatron, and the doughnut is small enough to be carried comfortably in a man's hands.

We can visualize the operation of the Mark I from what is known about particle accelerators. To begin with, high-energy electrons ionize the air surrounding them. This causes the betatrons to glow like an annular neon tube.

Therefore, around the rim of the saucer a ring of lights will glow like a string of shining beads at night. The power required for flight will ionize enough of the surrounding atmosphere to short out all electrical wiring in the vicinity unless it is specially shielded. In theory, the top speed of the Mark I is close to the speed of light; in practice, there are many more problems to be solved before relativistic speeds can be approached.

The peculiar property of microwaves heating all material containing the water molecule means that any animal luckless enough to be nearby may be cooked from the inside out; vegetation will be scorched where a saucer lands; and rocks containing water of crystallization will be blasted. Every housewife with a microwave oven knows all this; only hardheaded scientists and softheaded true believers are completely dumbfounded. The UFOnauts would be cooked by their own engines, too, if they left the flight deck without shielding. This probably explains why a pair of UFOnauts, in a widely published photograph, wear reflective plastic jumpsuits. Mounting the betatrons outboard on a disc is an efficient way to get them away from the crew's compartment, and the plating of the hull shields the interior. At high accelerations, increasing amounts of power are transformed into radiation, making the centrifugal drive inefficient in strong gravitational fields. The most practical employment of this engineering is for large spacecraft, never intended to land. The flying saucers we see are very likely scouting craft sent from mother ships moored in orbit. For brief periods of operation, the heavy fuel consumption of the Mark I can be tolerated, along with radiation leakage—especially when the planet being scouted is not your own.

When you compare the known operating features of particle centrifuges with the eyewitness testimony, it is fairly evident that any expert claiming flying saucers to be utterly beyond any human explanation is not doing his homework, and he should be reexamined for his professional license.

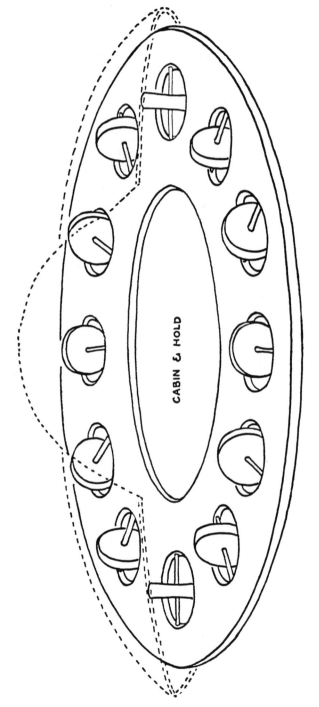

SCHEMATIC DIAGRAM
MARK I FLYING SAUCER

A SERIES OF ELECTRONIC CENTRIFUGES BASED ON THE HYPERSPACE OR PLANETARY GEAR DRIVES AND MOUNTED OUTBOARD ON A COWLED DISC ESTABLISHES THE CHARACTERISTIC PROFILE OF THE FLYING SAUCER.

For dramatic purpose, I have classified the development of the Flying Saucer through five stages:

Mark I—Electronic centrifuges mounted around a fixed disc, outboard.

Mark II—Electronic centrifuges mounted outboard around a rotating disc.

Mark III—Electronic centrifuges mounted outboard around rotating disc, period of cycles tuned to harmonize with ley lines, for jet assist.

Mark IV—Particle centrifuge tuned to modify time coordinates by faster-than-light travel.

Mark V—No centrifuge. Solid state coils and crystal harmonics transforms ambient field directly for dematerialization and rematerialization at destinations in time and space.

Now that the UFO phenomenon has been demystified and reduced to human ken, we can proceed to prove the theory. If your resources are like those of the PLO, you can go ahead and build your own flying saucer without any further information from me, but I have nothing to work with except the junk I can find around the house.

I found an old electric motor that had burned out, but still had a few more turns left in it. I drilled a hole through the driving axle so that an eight-inch bar would slide freely through it. I mounted the motor on a chassis so that the sliding bar would rotate in an eccentric cam. In this way, one end of the bar was always extended in the same direction while the other was always pressed into the driving axle. As both ends had the same angular velocity at all times, the end extending out from the axle always had a higher angular momentum. This resulted in a concentration of centrifugal acceleration in one direction. When I plugged in the motor, the sight of my brainchild lurching ahead—unsteadily, but in a constant direction—gave me a bigger thrill than my baptism of sex—lasted longer, too. But not much longer. In less than twenty seconds the burned-out motor gasped its last and died in a puff of smoke; the test run was broadcast on radio microphone but the spectacle was lost without television. Because my prototype did not survive long enough to make a run in two directions, I had to declare the test inconclusive because of mechanical

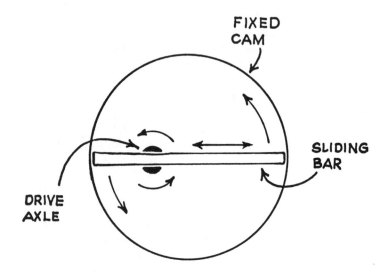

ESSENTIAL DESIGN OF THE HYPERSPACE DRIVE

THE ROTATIONAL VELOCITY OF THE OSCILLATOR IS CONSTANT ON BOTH SIDES OF THE DRIVE AXLE. THEREFORE, CENTRIFUGAL ACCELERATION MUST ALWAYS BE HIGHER ON THE LONG EXTENSION.

breakdown. So, what the hell, the Wright brothers didn't get far off the ground the first time they tried, either. Now that I know the critter will move, it is worthwhile to put a few bucks into a new motor, install a clutch, and gear the transmission down. One problem at a time is the way it goes.

A rectified centrifuge small enough to hold in one hand and powered by solar cells, based on my design, could be manufactured for about fifty dollars (depending on production run and competitive bids). Installed in Skylab, it would be sufficient to keep the craft in orbit indefinitely. A larger Hyperspace Drive (as I call this particular design) will provide a small but constant acceleration for interplanetary spacecraft that would accumulate practical velocities over runs of several days.

It is rumored that a gentleman by the name of Dean invented another kind of antigravity engine sometime during the past fifty years, but I have been unable to track down any more information except that its design consists of wheels within wheels. A gentleman in Florida, Hans Schnebel, sent me a description of a machine he built and tested that is probably similar in principle to the Dean Drive. Essentially, a large rotating disc has a smaller rotating disc on one side of the main driving axle. The two wheels are geared together so that a weight mounted on the rim of the smaller wheel is always at the outside of the larger wheel during the same length of arc of each revolution, and always next to the main axle during the opposite arc. What happens is that the velocity of the weight is amplified by harmonic coincidence with the large rotor during one half of its period of revolution, and diminished during the other half cycle. This concentrates momentum in the same quarter continually, to rectify the centrifuge. The result is identical to my Hyperspace Drive, but it has the beauty of continuously rotating motion. Now, if the Dean Drive is made with a huge main rotor—like about thirty feet in diameter—there is enough room to mount a series of smaller wheels around the rim, set in gimbals for attitude control, and Mr. Dean has himself a Model T Flying Saucer requiring no license from the AEC.

In 1975 Professor Eric Laithwaite, Head of the De-

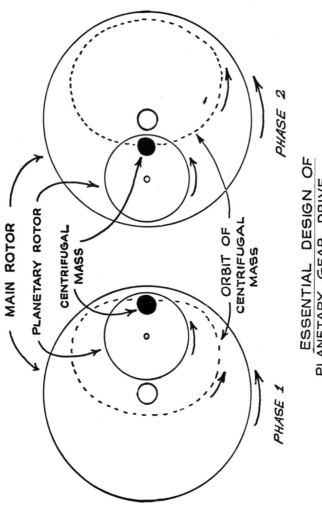

ESSENTIAL DESIGN OF
PLANETARY GEAR DRIVE

WHEN THE CENTRIFUGAL MASS IS AT THE RIM OF THE MAIN ROTOR, THE VELOCITY OF THE MASS IS ADDED TO THE VELOCITY OF THE RIM. WHEN THE CENTRIFUGAL MASS IS AT THE AXLE OF THE MAIN ROTOR, THE VELOCITY OF THE MASS IS SUBTRACTED FROM THE VELOCITY OF THE RIM. THEREFORE, CENTRIFUGAL ACCELERATION PREDOMINATES ON THE ONE SIDE.

partment of Electrical Engineering at the Imperial College of Science and Technology in London, England, invented another approach to harnessing the centrifugal force of a gyroscope to power an antigravity engine—well, he almost invented it, but he did not have the sense to hold onto success when he grasped it. Professor Laithwaite is world-renowned for his most creative solutions to the problems of magnetic-levitation-propulsion systems, and the fruit of his brain is operating today in Germany and Japan; his railway trains float in the air while traveling at over 300 miles per hour. If anyone can present the world with a proven and practical antigravity engine, it must be the professor.

Laithwaite satisfied himself that the precessional force causing a gyroscope to wobble has no reaction. This is a clear violation of Newton's Third Law of Motion *as generally conceived*. Laithwaite figured that if he could engage the precessional acceleration while the gyroscope wobbled in one direction and release the precession when it wobbled in other directions, he would be able to demonstrate to a forum of colleagues and critics at the college a rectified centrifuge that worked as a proper antigravity engine. His insight was sound but he did not work it out right. All he succeeded in demonstrating was a *separation between action and reaction*, and his engine did nothing but oscillate violently. Unfortunately, neither Laithwaite nor his critics were looking for a temporal separation between action and reaction, so the loophole he proved in Newton's Third Law was not noticed. Everyone was looking for action *without* reaction, so no one saw anything at all. Innumerable other inventors have constructed engines essentially identical to Laithwaite's, including a young high-school dropout who lived across the street from me.

Another invention is described in U. S. Patent disclosure number 3,653,269, granted to Richard Foster, a retired chemical engineer in Louisiana. Foster mounted his gyroscopes around the rim of a large rotor disc, like a two-cylinder flying saucer. Every time the rotor turns a half cycle, the precessional twist of the gyros in reaction generates a powerful force. During the half cycle when Foster's gyros were twisting in the desired direction, his clutch grabbed and transmitted the power

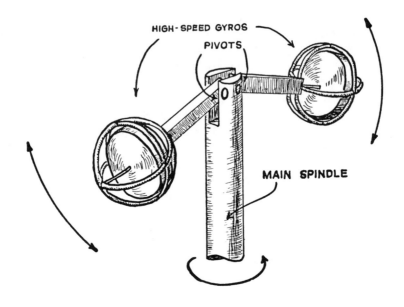

ESSENTIAL DESIGN OF THE LAITHWAITE ENGINE

WHEN THE MAIN SPINDLE ROTATES, PRECESSIONAL ACCELERATION CAUSES THE GYROSCOPES TO RISE AND FALL DURING EACH REVOLUTION. THE MECHANICAL PROBLEM IS TO ENGAGE THE RISE OF THE GYROSCOPES TO GENERATE A LIFT WHILE DISENGAGING THE DOWNWARD SWING.

to the driving wheels. During the other half cycle, the gyros twisted freely. Foster claimed his machine traveled four miles per hour before it flew to pieces from centrifugal forces. After examining the patents, I agreed that it looked like it would work, and it certainly would fly to pieces because the bearing mounts were not nearly strong enough to contain the powerful twisting forces his machine generated. Foster's design, however, cannot be included among antigravity engines because it would not operate off the ground. He never claimed it would, and Foster always described his invention truthfully as nothing more than an implementation of the fourth principle of locomotion.

What Laithwaite needed was another rotary component, like the Dean Drive, geared to his engine's oscillations so that they would always be turned to drive in the same direction. As it happens, an Italian by the name of Todeschini recently secured a patent on this idea, and his working model is said to be attracting the interest of European engineers.

When the final rectifying device is added to the essential Laithwaite design, all the moving parts generate the vectors of a vortex, and the velocity generated is the axial thrust of the vortex. Therefore I call inventions based on this design the Vortex Drive.

By replacing the Hyperspace modules of the Mark I Flying Saucer with Vortex modules, still retaining the essential betatron as the centrifuge, performance is improved for the Mark II. To begin with, drive is generated only when the main rotor is revolving, so the saucer can be parked with the motor running. This eliminates the agonizing doubt we all suffered when the Lunar Landers were about to blast off to rejoin the Command Capsule: Will the engine start? This would explain why the ring of lights around the rim of a saucer is said to begin to revolve immediately prior to lift-off. A precessional drive affords a wider range of control, and the responses are more stable than a direct centrifuge. But the most interesting improvement is the result of the *structure* of the electromagnetic field generated by the Vortex Drive. By amplifying and diminishing certain vectors harmonically, the Mark III Flying Saucer can ride the electromagnetic currents of the Earth's

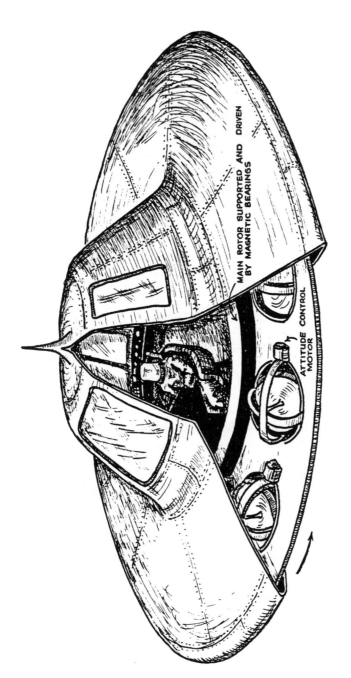

MARK II FLYING SAUCER

ELECTRONIC CENTRIFUGES BASED ON THE VORTEX DRIVE ARE MOUNTED IN GIMBALS TO TURN IN SYNCH WITH THE REVOLUTIONS OF THE MAIN ROTOR DISC.

THE TUNED ELECTROMAGNETIC FIELD GENERATED BY THE VORTEX DRIVE CAUSES THE VEHICLE TO BE CARRIED BY THE EARTH'S ELECTROMAGNETIC FIELD LIKE A DIRIGIBLE ELECTRON. CONTROLLED GEOMAGNETIC PROPULSION IMPROVES THE DESIGN EFFICIENCY TO THE MARK III STAGE

electromagnetic field, like the jet stream. And this is just what we see UFOs doing, don't we, as they are reported running their regular flight corridors during the biennial tourist season. Professor Laithwaite got all this together when he conceived of his antigravity engine as a practical application of his theory of "rivers of energy coursing through space"; he just could not get it off the drawing board the first time.

The flying saucer consumes fuel at a rate that cannot be supplied by all the wells in Arabia. Therefore we have to assume that UFO engineers must have developed a practical and compact atomic fusion reactor. But once the Mark III is perfected, another fuel supply becomes attainable, and no other is so practical for flying saucers. The Moray Valve will draw all the energy a saucer needs from the space through which it flies. The Moray Valve converts the Mark III into a Mark IV Flying Saucer by extending its operating capabilities through *time* as well as space. The Moray Valve, you see, functions by changing the direction of the flow of energy in the sun's gravitational field. It is the velocity of energy that determines motion, and motion determines the flow of time. We shall continue the engineering of flying saucers in the following essays.

My investigation into antigravity engineering brought me a technical report while this typescript was in preparation. Dr. Mason Rose, President of the University for Social Research, published a paper describing the discoveries of Dr. Paul Alfred Biefeld, astronomer and physicist at the California Institute for Advanced Studies, and his assistant, Townsend Brown. In 1923 Biefeld discovered that a heavily charged electrical condensor moved toward its positive pole when suspended in a gravitational field. He assigned Brown to study the effect as a research project. A series of experiments showed Brown that the most efficient shape for a field-propelled condensor was a disc with a central dome. In 1926 Townsend Brown published his paper describing all the construction features and flight characteristics of a flying saucer, conforming to the testimony of the first flight witnessed over Mount Rainier twenty-one years later and corroborated by thousands of witnesses ever since. (The Biefeld-Brown Effect explains why a Mark III rides the electromagnetic jet stream.)

We may speculate that flying saucers spotted from time to time may not only include visitors from other planets and travelers through time, but also fledglings from an unknown number of cuckoos' nests in secret experimental plants all over the world. The space program at Cape Canaveral may be nothing more than a supercolossal theater orchestrated by Cecil B. DeMille to reassure Americans that they are still *número uno* after Russia beat our atomic ace by putting Sputnik into orbit. We need not doubt that the Apollo spaceships got to the Moon, but we may wonder if Neil Armstrong was the first man to land there. The real space program may have been conducted in secret as a spin-off from the Manhattan Project since the end of World War II, and Apollo 13 may have been picked up by a sag wagon to make sure our team scored a home run every time they went to bat. The exploration of space is the most dangerous enterprise ever taken on by a living species. Don't you ever wonder why the Russians are losing men in space like a safari being decimated in headhunter country, while nothing ever happens to our boys except accidents during ground training?

From the book **How To Build A Flying Saucer** by T.B. Pawlicki, copyright 1981 by T.B. Pawlicki. Published by Prentice Hall, Inc. Englewood Cliffs, NJ. 07632

THE ANTI-GRAVITY EQUATION

Bruce L. Cathie
&
Peter N. Temm

Space travel, thanks to the American and Russian rocket programs, is a reality. By using great masses of hardware, miles of wiring and plumbing, Newton's law of motion and some not inconsiderable beginner's luck, the first man on the moon has stepped out of the pages of Jules Verne into history. Probes have been sent out to study Venus and Mars at close quarters. Considering the brief time in which man has taken serious notice of ways to bridge the gap between earth and sky, these are remarkable achievements.

Less than thirty years ago German scientists were carrying out experiments in rocketry; their experiences gained in directing doodlebugs toward London made them the leaders in this field; Werner Von Braun and other German experts subsequently passed on their knowledge and experience to others on both sides of the Iron Curtain. Those Model T's of rocketry, the V2s, have given rise to the sophisticated Apollo and Vostok power units which thrust comparatively tiny payloads into the void of space.

They work. And they have enabled man to venture far away from his own planet.

Yet, before the first space vehicle built on this planet thrust its snout beyond the ultimate atom of air in the envelope which clings as a life-sustaining wrapper about our planet, the method was already out of date. It may have been out of date centuries ago. As a means of propulsion rockets are obsolete-as obsolete as a propulsion method as a slingshot is out of date as a weapon. In both cases brute force is the criterion; in both cases the law of diminishing returns rules the game. As the power of the prime mover is increased, the size and cost of the system becomes less and less practicable, and the payload must eventually fail to give sufficient returns in relation to the effort expended in moving it through space.

A rocket allowed mankind to take the first essential step; but it is obvious, if we stop to think about it, that rocket propulsion cannot be the answer to exploration of the star systems that surround our galaxy.

Even if we could eventually achieve speeds approaching that of light itself, we could never hope to venture far from the solar system in one lifetime. The distances are vast; rockets the size of small cities would be necessary, so that several generations could live out their lives during the journey through space in order that there would be colonists for the far planet destinations.

It might be possible to do this. It would not be very practicable. As a theme for science fiction writers it has been a bountiful source of stories.

The first glimmerings of how true space travel might be achieved came to me when I uncovered the first clues that led me to the UFO grid which laces about our globe. I published my findings and hoped that someone with greater scientific knowledge than I possess would carry on from that point - and finally discover every secret locked into the system.

I was aware that my calculations were not precisely accurate - in the strict mathematical sense - but I could see that the system was based on space-time geometrics, and at least there was the best possible support for this: none less than the theories of Einstein.

Somewhere, I knew, the system contained a clue to the truth of the unified field which, he had postulated, permeates all of existence. I didn't know at the time that this clue had already been found by scientists who were well ahead of me in the play. I know now that they must have understood something of the grid system years ago. They knew that Einstein's ideas about the unified field were correct. What's more, for many years they have been carrying out full-scale research into the practical applications of the mathematical concept contained in that theory.

We are told that Einstein died without completing his equations related to the unified field theory. But in recent times it has been said that he did in fact complete his work; but that the concepts were so advanced that the full truth was not released.

I believe, and with good reason, that the full facts have been known to atleast some scientists for at least twenty years - and maybe much longer.

The very fact that man-made equipment has been built into the UFO grid system, which I put forward as a theory four years ago, is evidence that at least some people had prior knowledge of the grid and its workings. It must have taken them years to build up the network of stations which now exists. But as far as they or their masters are concerned, it is not *our* right to know.

The only way to traverse the vast distances of space is to possess the means of manipulating, or altering, the very structure of space itself; altering the space-time geometric matrix, which to us provides the illusion of form and distance. The method of achieving this lies in the alteration of frequencies controlling the matter-anti-matter cycles which govern our awareness or perception of position in the space-time structure. Time itself is a geometric, just as Einstein postulated; if time can be altered, then the whole universe is waiting for us to come and explore its nooks and crannies.

In the blink of an eye we could cross colossal distances; for distance is an illusion. The only thing keeping places apart in space is time. If it were possible to move from one position to another in space, in an infinitely small amount of time, or 'zero time', then both the positions would co-exist, according to our awareness. By speeding up the geometric of time we will be able to bring distant places within close proximity. This is the secret of UFOs - they travel by means of altering the spatial dimensions around them and repositioning in space-time.

When I had completed my first book I was not aware of just how close I had come to this truth. The answer was literally staring at me from the pages of my own work for nearly three years, before a visit from a stranger brought it into my grasp.

A year after the book was published I had a telephone call from a man who had just arrived in New Zealand from England. He explained that he was a textile salesman and would be here only a few days on business. he said he had heard of my research, and was most insistent that he should see me. He said that he had very little knowledge of UFOs, but that he would like to talk to me about my theories. Could he come out to my home for a visit?

He arrived in a rented car, and at once began questioning me about my activities and my research. It was soon obvious that he knew a great deal more about the subject of UFOs than he was prepared to admit; and he was quite demanding about

"Mystery Spot", Santa Cruz, California—a gravity-anomaly area. Note the apparent change in height of the two men when they exchange positions.

getting his questions answered. He had an air of nervous tension about him as he checked through pages of my calculations; then he wanted to know where I kept all my data, and if there were many people who knew about my studies.

I make no secret of whatever findings I turn up, and I showed him everything he asked to see. Finally he insisted that there was something I hadn't shown him - an equation which knitted all my calculations together. In some surprise I told him I knew of no such equation; his expression was eloquent of disbelief.

As this discussion proceeded I informed him that there were probably others who thought as he did - that I had some equation up my sleeve, because I was being watched and on several occasions my car had been followed. I added that since he had come to my home he would most likely have unwanted company on his way back to the city. Whoever it was that was keeping an eye on me, I told him, would probably believe him to be a contact man and therefore would be interested in his movements.

I must admit that I told him this intriguing little tale to see what his reaction might be. I was fascinated to see his apparent nervousness become increased, and in some agitation he decided it was time to leave. His departure was hurried; I have neither seen him nor heard of him since.

For the next two or three days I found myself going over and over his questions. The more I thought about it, the more convinced I became that the stranger had known far more about my calculations than I did myself. Was there something there which he could see - but which I had missed? If there was an equation buried somewhere in my figuring, then I would have to find it.

Reworking the math, I finally decided to concentrate specifically on three harmonic values which appeared to have a close relationship with one another. Previously, I had shown this connection, and had truthfully pointed out that I did not know why the relationship was there at all. These were the harmonic values that now fully occupied my attention: 1703 - this is the four-figure harmonic of 170,300,000,000, which is the expression in cubic minutes of arc of the mass or volume of the planet earth and its surrounding atmosphere. 1439 - A four-figure harmonic of 143,900 minutes of arc per grid second, representing the speed of light in grid values. 2640 - This figure, expressed in minutes of arc values, is built into the polar portion of the grid structure as a geometric co-ordinate. (The polar diagram contained in Harmonic 33 is reproduced at the back of this book, and illustrates this value. At the time I wrote my earlier book I was not aware of what this value represented.)

Now I found that when I matched these values harmonically the results were as follows. Zeros to the right-hand side can be ignored in this form of harmonic calculation:

$$\begin{array}{r} 1703 \\ -264 \\ \hline 1439 \end{array}$$

In other words the difference between the harmonic of mass and the harmonic of light is the harmonic of 264 (or 2640). It was now apparent that if my calculations were more accurately worked out it should be possible to find out just what the 2640 figure referred to.

After several hours of work the following was what looked up at me from my paper:

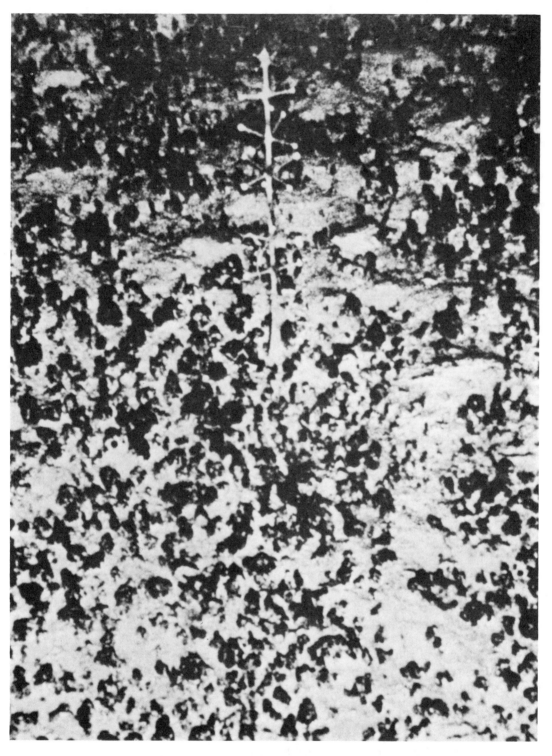

An "OOPART". Photographed on the sea-bed at a depth of 2,500 fathoms, 1,000 miles west of Cape Horn, by US survey ship *Eltanin*, whose officers are close-lipped about it. "It's a marine organism," maintained an *Eltanin* officer in Auckland in 1968. Pressed further, he admitted: "But it still looks like an artifact to me!"

17025 earth mass harmonic
-2636 unknown harmonic
───────
14389

    Checking through the five-figure mathematical tables I found to my surprise that 2.6363 is the square root of 6.95 (from the 1-10 square root tables). In harmonic calculations of this kind decimal points as well as zeros to the right of a figure can be ignored; so it could be said that the sqare root of 695 is 2636. Now I could percieve the first steps necessary to solving the elusive equation. I had long established that 695 is the harmonic reciprocal of the speed of light, or 1/1439. The calculations were now sufficiently accurate for algebraic values to be substituted - although obviously a computer would be necessary to solve the true values to extreme accuracy.

So -

17025 (earth mass)
-2636 (square root of speed of light reciprocal)
───────
14389 (speed of light)

If C = the speed of light, and
M = mass

Then $(C + \sqrt{\frac{1}{C}}) = M$

    Now at least I had the first part of a unified field equation in harmonic values. To take the next step I first had to go back to the Einsteinian theory, particularly the famous equation, $E=MC^2$, where E is energy, M is mass and C the speed of light.
    Einstein declared that physical matter was nothing more than a concentrated field of force. What we term physical substance is in reality an intangible concentration of wave forms. Different combinations and structural patterns of waves unite to form the myriads of chemicals and elements which in turn react with one another to form physical substances. Different wave forms of matter appear to us to be solid because we are constituted of similar wave forms which resonate within a clearly defined range of frequencies - and which control the physical processes of our limited world.

    Einstein believed that M, the value for mass in the equation, could eventually be removed, and a value substituted that would express the physical in the form of pure energy. In other words, by substituting for M, a unified field equation should result which would express in mathematical terms the whole of existence - this universe, and everything within it. As I have already said, it seems that before his death Einstein did indeed produce this equation. What the mathematical terms were which he used, I do not know; but I couldn't help wondering what results I might get if I were to tackle the problem from the point of harmonic math. If an equation could be found which had a harmonic affinity for all substance, one which showed a resonance factor tuned to matter itself, then the problem, perhaps, would be solved.

Einstein maintained that the M in his equation could be replaced by a term denoting wave form. I had now found a substitute for M in terms of wave form of light. So the obvious step, to me, was to replace Einstein's M with the values of C, found on the UFO grid. These are the results I obtained:

$$\text{Einstein: } E = MC^2$$

$$\text{UFO grid: } M = \left(C + \sqrt{\frac{1}{C}}\right)$$

$$\text{Therefore: } E = \left(C + \sqrt{\frac{1}{C}}\right)C^2$$

I now had right before me a harmonic field equation expressed in terms of light - or pure electromagnetic wave form - the key to the universe, the whole of existence; to the seen and the unseen, to form, solids, liquids, gases, the stars and the blackness of space itself, all consisting of visible and invisible waves of light. All creation is light; that was the answer that had been right within my grasp for four years.

Now it was necessary to check the validity of the equation; to accomplish this I needed a computer and the help of an expert in pure mathematics.

Harvey Patchet is a young man who had attended one of my lectures. He said he was willing to put a small computer at my service, and a friend of his, an excellent mathematician who had a larger computer, was willing to program it to give me assistance should I need it. I was deeply appreciative of both offers; throughout all my research I have always been at the limits of my own knowledge, forever frustrated by my slow rate of progress. I could always see much further ahead than my technical ability would allow me to travel.

In the earlier editions of this book I had believed thatt the value of 2545.56, the geometric distance in minutes of arc, from the grid poles to the main corner aerial positions, in the grid polar squares, was a harmonic of gravity acceleration. I am now aware that this assumption was incorrect, and that the true value for the harmonic of gravity acelleration is 28.24 grid feet/grid sec^2. A full explanation for the derivation of this figure will be found in my third book, **The Pulse of The Universe, Harmonic 288**.

The reciprocal of 2545.56, rounded off to a four-figure accuracy, was given as a harmonic of 3930. When I made my initial discovery of the unified equations, the harmonic of 3930 was thought to be that of anti-gravity. It is now known that this value is related to the earth's magnetic field. This is also explained in my third book. Although I labeled them wrongly, the actual values have stood the test of time, and still fit into the equations as originally demonstrated.

The basic equations shown in this chapter are fundamentally correct, but have been extended by the insertion of the gravity factors in my later work.

The equations, using harmonics of single C values, are those from which an atomic bomb is developed. By setting up derivatives of the equation in geometric form, the relative motions of the wave forms inherent in matter are zeroed, and convert from material substance back into pure energy. By reversing the process, physical substance in any desired shape or form could be produced from pure energy.

While practical applications of the latter possibility may still be far in the future, we are nevertheless at least capable, now, of destroying matter. As I have stated elsewhere - scientists can make a bang from a bomb, but as yet they are still unable to get the bang back into the bottle.

Although I was much happier with progress when I arrived at this point in my calculations, there were still some important answers missing. As far as they went, the equations explained the workings of a nuclear explosive device, but they had not yielded the secrets of UFO propulsion. The UFO grid was undoubtedly constructed on the basis of the equation, yet a UFO does not disintegrate when it moves within the resonating fields of the system. There had to be an extension of the equation, harmonics for movement in space-time. But I felt close to the solution, and I was determined to find it.

It has often been my experience that when a stalemate is reached it is best to throw aside one's work for a time and take a rest. For the next few weeks I deliberately forgot everything evenly remotely connected with UFOs. But one evening, unheralded and unexpected, a thought flashed into my mind. Why only deal with single values of C? The key was turned and a sequence of doors swung open, one after another, to reveal secrets of space and time.

In the polar area of the UFO grid the geometric values of some of the co-ordinates appear to be doubled up.

The diagrams of the 'polar squares', as I have termed them, incorporate twice the value of 2545.56.

In the resultant grid polar square the speed of light reciprocal harmonic of 695 is incorporated in a harmonic of 2695. These co-ordinates are the same from corner aerial positions to the geographic poles. It appears that the factor 2 preceding the 695 serves to harmonically double the reciprocal of the speed of light.

I reasoned that a way to check this idea was to increase the values of C in my equation, and oserve the changing harmonic of E. In a few hours in persuit of this line of thought, I was able to say: 'I have found something extremely interesting'.

It was apparent that one only had to do double the value of C. Once I had done this, it was possible to produce two different equations, through the use of tables relating to the square root portion of the equation. The results differed according to whether square roots 1-10 or 1-100 were employed. In equation No.1 I found the resulting harmonic was 3926991712050.

This was the value calculated by the computer after we had fed the equation in, and gradually corrected from one side of the equation to the other. I was only too well aware that my calculations were only of 'near-enough' accuracy up to this point, and had been sufficient only to form a general basis for the theories I had developed. Four-figure accuracy in fact is just not nearly good enough, as any pure scientist will be quick to agree. By feeding in the harmonics on either side of the equation and gradually correcting, the 12-digit value was finally arrived at.

This harmonic is not yet completely accurate; larger computers will be necessary to carry on the correction to the ultimate of precision. Nevertheless, the result is now 'good enough' as far as I am concerned, and it may at least start off a scientist of an independent nature to carry out further investigations of his own.

I had shown previously that the four-figure harmonic of 3930 (rounded off to four figures) was the reciprocal of 2545.56. In a practical sense the values were

What caused these marks in manuka scrub at Ngatea, New Zealand, in 1969? None of the official "explanations" makes sense.

sufficiently accurate to demonstrate what I was trying to establish. Therefore, I considered that the harmonic of 392699171205 derived by doubling the C component of the equation had to be the reciprocal of the 2545.56 value found in the grid polar square areas. The equation could now be expressed thus:

$$\frac{1}{2545.56} = \left(2C + \sqrt{\frac{1}{2C}}\right)(2C)^2$$

It is interesting, even curious, to note that the two main stations in New Zealand manned by overseas scientists - namely, at Woodburne, in the South Island, and Kauri Point near Auckland - are spaced at a geometric distance of 2C minutes of arc.

The second equation is derived by using the square root of the 10-100 tables. This equation is as follows:

$$\text{Double harmonic of the speed of light reciprocal} = \sqrt{\left(2C + \sqrt{\frac{1}{2C}}\right)(2C)^2}$$

The nearest we have brought this to accuracy, using a small desk top computer, is as follows:

$$2695 = \sqrt{\left(2C + \sqrt{\frac{1}{2C}}\right)(2C)^2}$$

$$= \sqrt{\left(28777620 + \sqrt{\frac{1}{28777620}}\right) 28777620^2}$$

$$= \sqrt{(28777620 + \sqrt{34749200})\, 82815141}$$

$$= \sqrt{(28777620 + 58948452)\, 82815141}$$

$$= \sqrt{87726072 \times 82815141}$$

$$= \sqrt{72650740}$$

$$= 26953$$

The speed of light harmonic accurate to two figures is equal to 1439, as I have said elsewhere. The value used in this calculation according to the more accurate computer check equals 1438881. As I have already stated, the true value will only be derived by the use of highly sophisticated computers.

Another possible lead for mathematician readers is the fact that the natural tangent of 36° is .72654. The sqare of 26953 from the equation is 726504, which is very close.

The fact that the harmonic of light has only to be doubled to obtain anti-light fields must be related to the matter and anti-matter cycles of the physical and the non-physical worlds. If the two, plus and minus, fields are interlocked, as I have postulated, and matter and anti-matter manifest in alternate pulses, then a double cycle must occur between each pulse of matter and anti-matter. The anti-matter pulse cannot be percieved by us, for fairly obvious reasons; but when calculating the frequency interaction between the two, both cycles must be taken into account.

By stepping up or slowing down the frequency of C between the two cycles, a shift in space-time **must** occur.

Copyright ©1971 by B.L. Cathie and P.N. Temm. From, "Harmonic 695, The UFO and Anti-Gravity" BY Bruce L. Cathie and Peter N. Temm. All rights reserved by Quark Enterprises Ltd. 158 Shaw Rd., Oratia, Auckland, New Zealand.

THE AUTHORS

CAPTAIN BRUCE CATHIE is a Fokker Friendship and Boeing 737 captain with the National Airways Corporation of New Zealand. After he and a number of other flying enthusiasts had sighted a UFO at Mangere, Auckland, in 1952, Captain Cathie began to apply his professional knowledge of navigation theory to the UFO mystery.

His discovery of a world magnetic grid system was proved in his first book, *Harmonic 33*, published in 1968, which has become famous in UFO circles throughout the world. His researches since its publication have led him on to the still more astonishing discoveries detailed in *Harmonic 695*. Captain Cathie is married with two children and lives in Oratia near Auckland.

PETER TEMM, who died in 1976, was a correspondent for newspapers including the *Washington Star*, the *Sunday Times* (London), the *Melbourne Herald* and the *Auckland Star*. His interest in UFOs began when he was in Japan studying the "sky myths" of the Ainu people.

Up In The Air Over Anti-Gravity ???

By W. P. Donavan

Disk, Disk

A discussion on Anti-Gravity research would naturally begin with an introduction to the pioneering work of Nikola Tesla, John Searle, Bruce DePalma, T. Townsend Brown and Thomas Bearden. Let's start by looking over the illustrations concerning the Searle Disk and DePalma's "N" machine. A close examination will reveal a few similarities between the "N" machine and the Searle Disk. Keep that thought in mind.

A brief explanation is in order. After reading the excerpt by Bruce DePalma beneath the illustration of his "N" machine, you might be wondering **how** you can get a potential across practically nothing. If you embrace the old theories on the structure of mass, energy and space, that would be an obvious impossibility. And it would be if space is regarded as a static medium. That is the essence of the theory--that space, instead of being "dead", is a dynamic medium alive with activity. This is the philosophy that Nikola Tesla used to construct one of the first free energy machines. I would heartily recommend reading any information that the inventor had written in his own hand on the subject. His writings, even though they're pretty cryptic, contain the key.

In <u>Ether Technology</u> by Rho Sigma (see bib.), the view is put forth that gravity is electrical in nature, and in Thomas Bearden's works as well (<u>Toward a New Electro-magnetics 1,2,3,& 4 see bib.</u>). Jerry Gallimore has done extensive research toward that end also (you'll see one of his papers in the proceedings of The <u>First International Symposium On Non-Conventional Energy Technology</u>. (See bib.)) The basis of the "N" effect is that when a stress in the form of a centrifugal force (which is actually a form of gravity anyway) is placed on a material in a magnetic field which is at right angles to the force, an electrical potential is present. Now when a material is in a magnetic field a measure of coherence in atomic motion is conveyed to that material. In the absence of a magnetic field the material of the rotor will have a fraction of the voltage present. This clearly demonstrates that that which is termed, the "motional electric field" (Dr. W.J. Hooper professor at UCLA, Berkley first proposed the term), is intrinsically linked to the phenomenon of gravity. One is never seen without the other.

Therefore, it should be possible to produce an electric field with all the properties of a gravitational force field. All it takes is a little ingenuity. Let's see... You would need an extremely high voltage if you had a material with a comparatively low gravitational coupling coefficient. Or vice-versa. What does that suggest to you? It seems to me that a parallel-plate capacitor has the same ability to transform an electric field into a gravitational field that an electromagnet has in the transformation of an electric field to a magnetic field. And between the two lies an inextricable link, the electric field. If there is any point where I would start in the search of a unified field theory, that would be it.

The Searle effect is the same as the "N" effect. In the case of the Searle disk the magnetic field poles are at right angles to the axis of rotation, while in the "N" machine the magnetic field poles are parallel to the axis of rotation. The electric field that is generated is parallel to the spin axis on the Searle disk. The Biefield-Brown effect is also linked because the gravitational coupling coefficient is critical to the operation of the device. But the Searle Disk seems to obviate that problem by using the brute force approach and using a voltage with an extremely high value. So the Searle effect uses the Biefield-Brown effect in its operation.

One thing that bothered me about the Searle disk is that it so closely resembled the "N" machine of Bruce DePalma. This is the result of some deep probing into that nagging hunch that seems to have validated the claims made by J. R. R. Searle.

Experimental set-up of the British inventor John Searl

Look at the "N" Machine. Notice that the electric field vectors appear at right angles to the magnetic field vectors?

Before I go further it would be prudent to mention something about a bit of trivia that the experimenters choose to ignore... probably because they can't explain it. If you measure the electric field potential with a "fixed" voltmeter on the rotating disk you will notice a very real voltage. But if the voltmeter is mounted on the rotating disk it will show absolutely no deflection. Why is that? There is an explanation for that but it involves some esoteric wave mechanics. Basically the wave function of the electron collapses on the rotating reference and becomes a probabilistic function. In other words, the particle vectors and the quantum field 'smear' into each other so as to be indistinguishable from each other. You could have an infinite voltage on the disk and it would not arc over onto another portion of that disk. Surprise! That's exactly what the Searle disk does.

Put the drawings of the Searle disk and the "N" machine next to each other and look at them. What similarities do you see? Notice that the magnetic poles in the "N" machine are parallel to the axis of rotation while the poles on the Searle disk are at right angles. Now if the electric field vectors are at right angles to the magnetic pole orientation, that means that the electric field vectors

The "N" Effect

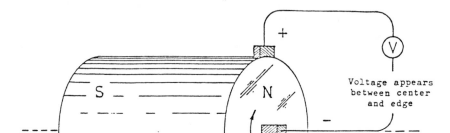

Rotating Permanent Magnet (Alnico)

The N Machine

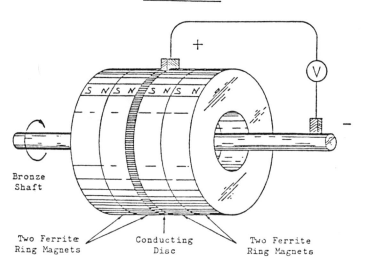

Two Ferrite Ring Magnets   Conducting Disc   Two Ferrite Ring Magnets

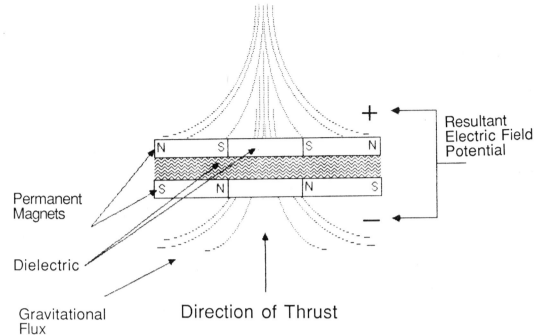

Field Configuration of Searle Disk with modified hydrodynamic vortex

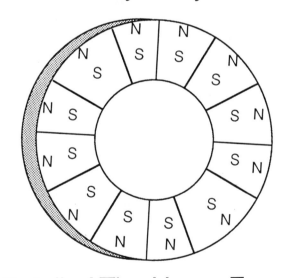

Detail of The Upper Face Showing The Magnet Positions

---- Diagram G ----

are parallel to the spin axis of the disk. And if your magnets are the crystalline variety, that makes the disk a capacitor with an unbelievable value! Now if the formulas for the Biefield-Brown effect are valid, you'd get an electrogravity field proportional to the potential across the disk. Of course since the electrons are a probabilistic function <u>on</u> the disk, there's no possibility of arcover and no dielectric stresses in the material.

That means that you could have 10^{15} volts across the disk measured from a "fixed" reference and zero volts on the disk itself without any danger of it becoming a bomb. Another interesting point suggests itself. You could also have an infinite charge on the disk surfaces without any leakage to the adjacent plates. Infinite energy on the outside, zero energy on the inside. How's that for a model of a U.F.O.? It would also have an integral force field operating on the outside of the disk. How, you may ask?? Very easy. Once the disk hits the right speed the scalar electrostatic field zipping around the periphery reaches the potential that quenches electron conduction and generates a "skin" effect analogous to barrier turbulence in an aircraft. Any mass approaching is first attracted and then violently repelled. So you have a deflection shield a' la' Star Trek. Speaking of fields, what do you think would happen if you ran the current that it generated into a field coil located on the disk itself? Obviously from an onboard perspective there's no current flowing through the conductor and therefore no magnetic field results. But... from the outside you <u>do</u> have a current flow. And that does produce a magnetic field. So, on the disk you have a condition that bypasses Lenz's law and any need to generate a back E.M.F.

That's one point that bugged me to no end. Everybody wondered how Searle got around generating back E.M.F. on his machine when the "N" machine was doing exactly the same thing. Of course I overlooked it too. It's the obvious that's the hardest to see. Murphy's law.

Remember when I said that you could have an infinite charge on the disk surface that wouldn't be seen by the rotating disk? The charge would be seen on the outside as the scalar electrostatic potential, or S.E.P. for short. If you slowly cranked up the value of the S.E.P. , first the electrons would collapse into the quantum field. Then the protons, neutrons and the assorted zoo of subatomic particles would follow suit. The disk, and whatever was attached to it, would vanish into the quantum field. **Poof!** And it's gone. But what about the disk's perspective?

From the disk's point of view the S.E.P. never changed. It sees the universe as it always did, although now it exists solely in the quantum field. And that means it has a few extra dimensions of movement. That means anywhere in space, and anywhen in time. Since it exists as a probabilistic construct its existence is equally probable anywhere in the universe at the same time or anywhen in the universe in the same place. Shades of Doctor Who!

Travel in space would be easy enough. You just go wherever you want to go in the inertialess state and then switch off the field by de-spinning the rotor. If you wanted to, travel could be instantaneous. The difficulty would lie in time travel, which contains within itself dimensions of probability. It very well could be that the faster you go in space the more acute the curve becomes as far as a probability is concerned. So you'd travel through parallel worlds if you weren't careful and never return to where you'd started from. If anybody out there is fooling around with this stuff, you have been forewarned. So there.

However, what I think happens is that the disk is undergoing a transcendent shift in its dimensional references. I should explain that a little more clearly. Remember in high school when they told you that when a particle is accellerated to the speed of light it becomes infinitely massive, so massive in fact that the entire universe would collapse on it from the resultant gravity field that it radiated? Well, in <u>The Excalibur Briefing,</u> (see bibliography) Tom Bearden clears up that absurd mess. What really happens is that the particle loses a dimension and becomes two-dimensional. And at the same time it loses mass and has infinite momentum, or inertia. Their big mistake was equating inertia and mass. (A good example of that is the photon. It has no mass and cannot be accellerated beyond its inherent speed or slowed down to a point below it. In other words the particle has infinite momentum. So all real mass-energy conversion involves an explosive accelleration of mass to the speed of light.)

The conversion process of the material of the disk is another matter (pun intended). In that case it loses mass and spatial momentum, but has infinite temporal momentum. That would be an extreme case, however. In the real world, if you could call the quantum field that, it would be a value between zero and infinite. And your time vector would be dependent upon your temporal momentum and position on the lotus which is shown on diagram A. You've probably noticed by now what that implies. Your direction in time (future or past), is inversely proportional to mass and spatial momentum and proportional to temporal momentum. That determines your speed on the time line.

In that case how about turning the Searle disk into a full fledged time machine? In theory it would be possible. Instead of just one disk with the north poles facing out you use a paired set. The one on top has north poles facing out and the one beneath it has all of its south poles facing out. We know that the direction of spin, clockwise or counterclockwise, doesn't matter. The electric and gravitational field vectors are dependant upon magnetic pole orientation. With the two disks counterrotating at the same speed, and assuming both have the same field strength, the gravitational force vectors cancel out. In fact all the fields cancel out except for one. And that's the S.E.P.. The accumulation of that field would cause collapse of the wave function just as the field from the single disk did.

Now most people with a bias toward solid state operations would probably be recoiling in horror at the prospect of two huge counterrotating disks and the associated precessional and nutational torques contained within them. And rightly so. And the first thought that would come to their minds might be a configuration using a set of solenoid electromagnets with all of their north poles facing out mounted on the disk and sequentially feeding square wave pulses into them. I don't think that would work, and it's for this reason: The system is dependent upon an accellerated mass that has an integral magnetic field contained within it. But there is a way around that.

What does an accellerating body generate? It generates gravity waves! And so the problem becomes this: How do you generate the same type of gravitiational radiation that a rotating mass produces? Or, let me put that another way; how do you produce a "motional gravity field"? A motional gravity field would be the same as a motional electric field with a skewed vector. A look at the work of Hooper and Townsend Brown gives a great deal of insight on the matter. If you take a look at Townsend Brown's British patent no. 300,311 figures 6,7 and 11 it's obvious that's what he intended to do. But reading the entire patent gives a good backround on the matter. It illustrates that if you arrange the plates radially around a center and put a D.C. bias on them you'll get a field that looks like a moving mass. Now you can put the field coils on that disk and sequentially energize each one to make it look like a moving magnet set. That model might just work.

All this stuff with the strange effects of the Searle disk and the "N" machine makes you wonder what the hardware looked like aboard the U.S.S. Eldridge, the ship that played a part in the infamous Philadelphia Experiment. Of course back then they called it Project Rainbow. I'm convinced that whatever they could have used must have been based on the same effect as the Searle disk.

The original intent of "Project Rainbow" was an exercise in conservation. At that time copper was getting scarce, (this was about the same time that steel pennies were being minted) and huge coils that ran from bow to stern were required to degauss or demagnetize the ships' hull to prevent it from being destroyed by magnetic mines. Something else was needed to decrease the phenomenal consumption that each ship required to protect itself. Different coil geometries were tried. After numerous design changes the ultimate form was selected. This type apparently had a resemblance to a cadeuceus coil wound into a toroid configuration. A number of these coils were mounted around the interior of the hull. When they were energized, (according to the theory) the magnetic field in the hull would be quenched and thus eliminated. However, things didn't work out that way. When the coils were energized a field fundamentally different from what theory dictated was created. That field changed the amount of basic energy that the space contained and which the ship existed in. It increased it.

The Lotus

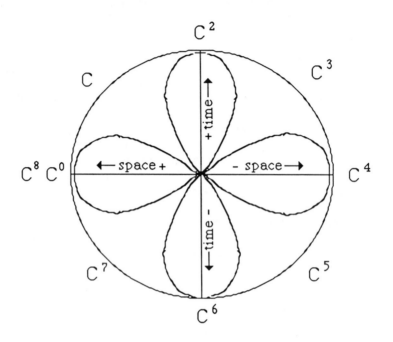

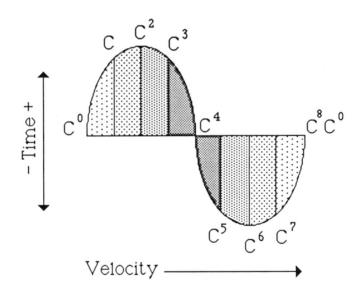

Diagram A

Now, the only reason that a material object can remain visible and tangible is for it to exist in a low energy-density per unit volume of space. If that energy density and charge were increased substantially, enough to equal an electron, then that electron would become indistinguishable from free space. If that energy density and charge were increased beyond the amount required to quench an electron, enough to quench a proton, then neither electron nor proton would exist as physical entities. A slight increase in the field from that point would cause complete dematerialization of the body.

Diagram A

Now that I've given you a few of the more preliminary concepts, it's time to dive headfirst into a deeper theoretical examination of the mechanisms. First we'll take a look at diagram 'A'. What it's about concerns some graphs of the velocity gradient. (I like that term; my thanks to Tom Pawlicki who first coined it.). Any particle or material body found in the physical universe occupies a position on this graph. The upper graph is what I term "the lotus", because it reminds me of the flower. Suffice it to say that any mass which is accellerated follows the outer curve for velocity and the inner, petal-shaped curve for momentum. The perimeter of the circle represents zero momentum or inertia, and the center represents infinite momentum or inertia.

Lotus Entertain You

The second part of diagram A shows what looks like a sine wave. This is a simplistic representation of a multidimensional plot. Here you see time flow (relative time flow) charted against velocity. C^2 would correspond to an infinite "timelike" velocity into the future and C^6 an infinite speed into the past. At the crossover points of C^0, C^4 & C^8 time ceases to exist for the accellerating body. One point becomes pretty obvious, and that point is that the curve as shown would be infinitely repetitive. And so is the lotus. The lotus would describe a spiral or corkscrew configuration which itself is curved. This curve would wrap itself into a toroid geometry. And that toroid is a part of a larger structure at right angles to itself. And so on, ad infinitum.

About that reference to the multidimensionality of the plot--it could conceivably be done on a computer plotting the variables, space, time, velocity, momentum, and skew angle on the lotus. When the body accellerates from C^0 to C^8 it describes a circle greater than 360 degrees. That is somewhat difficult to show in a two dimensional diagram.

One point of conjecture which I'd like to make which could turn out to be true or merely a lot of wasted ink, is this: the "T" core on the lotus, the point where the time, space and momentum axes meet and all three cease to exist would be synonymous with the quantum field. Here's another beastie to add to the quantum zoo. There's a particle that would fit rather neatly into the diagram. Call it an exo-tachyon. That particle would ride the hyperbolic section of the momentum curve like a surfer rides the waves. And quite possibly it could go in both directions too. Let's say you started on the portion of the curve between C^4 and C^5. The hyperbola would carry the particle to the point between C and C^2. It would then move to the point C^2-C^3 and through the curve to C^7-C^8. Then from just past C^8 down the curve to C^5-C^6, on to C^6-C^7 to C^3-C^4. Its antiparticle (?) would follow the curve in the opposite direction.

Conjugate exo-tachyon pairs would rotate about each other similar to the action of electrons and positrons in a supercooled state. (Conjugate electron-positron pairs are called positronium). There would be a good chance that particle "tunneling" would occur between layers of the spiral as well, and the crossover point would be at the "T" core. Of course that would mean that **most** of the particles would reside at the "T" core. And since whatever is in the "T" core has infinite momentum and zero mass that would also be one of the properties of the particle-also of free space, because it would be swarming with the stuff. Which would be entirely logical. After all, what can you do with free space? You can bend it, stretch it, change its geometry--but one thing you absolutely cannot do to it is accellerate or decelerate it. That is a definite no-no. If space had zero momentum there would be an infinite speed of physical objects with zero time flow. But what happens when you stop time? The physical properties of space change. It becomes impossible to acellerate a body. Because without time, space has infinite momentum. I think that what I term the "exo-tachyon sea" is analogous to the neutrino sea, zero-point energy, spacic energy sea, virtual particle sea, the ether, and whatever new terms have come round to describe it. In fact, it's also possibly the protomatter that Itzhak Bentov describes in <u>Stalking The Wild Pendulum</u>. (See Bibliography)

If it sounds like I'm getting too deep it's just that it's difficult to discuss a subject like free energy or gravity without getting a little heavy. Pun intended.

Now on to diagram A. Basically all it consists of is a table that charts out the properties of free space and time at the positions mentioned on the lotus. Speaking of the lotus--wherever the momentum curve converges on the "T" core the body posesses infinite momentum and zero mass. At the tip of the "leaves" you would have maximum mass and zero momentum. But now to the chart. At the position "C" the body would be in positive space, positive time and have infinite momentum. Those distortion effects are based upon Einsteinian space and may be wrong, but I'm using it anyway until I can prove it otherwise.

At position "C" the body drops a dimension and becomes two dimensional. At "C^2" the value for t becomes absolute and space has a null value. Also momentum has a null value. Since mass has a finite value from the perspective aboard the body but also occupies all space at the same time, the mass does not exist as a physical entity from our perspective. Of course it does exist, but all the way along the velocity gradient or continuum those realities become relative quantities and are by no means concrete or objective. Also at "C^2" time flow is infinitely fast. This is because the spacial dimension has collapsed into time. The next point, in my opinion, deserves some argument. That point is whether or not the body undergoes a dimensional transition. If I interpret Tom Bearden correctly, he believes that it does and that it becomes one dimensional and collapses into a line. I don't think that it does undergo a change and I'll outline my reasons why. My first objection is that it would completely screw up my theory! As an individual I <u>do</u> like order and accepting that particular element would throw a monkey wrench into the whole thing.

Second is that a dimensional transition at every point would lead to too many negative numbers, and intuitively it would not look right. But that of course is looking from a 3D perspective. If you begin with 11 dimensions then they'd all collapse into 3 in the end. And that would tie in with the new theories of dimensional expansion in the early stages of the universe. So what I'm going to do is give you my theory and let you decide which one sounds better.

At "C^2" the body does not undergo a dimensional change from an objective standpoint. At "C^3" you're still in positive time, but now you've got negative space to contend with. That makes for some very interesting physical properties. Negative impedance for one. Also negative permeability and permittivity for another. If you could get a mass into that section of the lotus and build a parallel plate capacitor with that space existing between it, then you'd have yourself a free energy device. Of course you would also need to find a way to scavenge the energy back to this section of space-time. But it could conceivably work.

At C^4 the mass rests right on the equator of the lotus. And 'rest' would be an apt description in this case, for the mass truly <u>is</u> at rest. It also has exploded into the quantum field (yech-what a mess!) and exists as an entity which is evenly distributed throughout the universe. Another property at this position is temporal simultaneity. The mass exists practically everywhere at the same time. Of course in reality it only exists during one particular collapse phase of the chronon. I'll explain the mechanics of the chronon at some other point since it becomes complicated very quickly.

At C^5 the mass exists in the flip side of the universe. All spatial and temporal values are reversed. Time moves backwards <u>and</u> space has negative values. In this part of the cosmic boondocks you'd probably find that hypothetical beastie 'negative matter'. Strange stuff! Since the inertial properties are reversed as well, if you pushed it the object would come toward you. If it were located in earths' gravity field the effects would be spectacular. For one thing any pull on the object would be interpreted as a push, so the mass would accellerate at 32 feet per second^2 away from the planet. It would literally fall up.

The thinnest slice of the material (and it does not depend on what material you use-- it's a condition of matter) would do. If you stood on a thin disk of the material, provided it had the necessary structural strength, two effects would occur which complement each other. The first is that effect which manifests itself in a gravity field (as if that weren't enough). The second is another one of its bizarre properties which makes a very useful propulsion system. Since the material is just as transmissive as normal matter to gravity waves you are attracted to the planet. But... you are also standing on the disk and applying force to it. So the disk is pushing back and negating any effects of the mass on the disk. This would produce what appears to be an inertial shield but which in reality is not. You would still feel the same g forces that the accelleration imposed on you as with any other form of conventional propulsion. The real question is how you would transform a mass, which for all practical purposes is a geometric point to our perspective into a real 3D object which could be used as a practical propulsion system. The answer is I don't know. At this point I only have vague conceptions on how to do it, and those will follow.

What you would need is a field which changes the structure of free space that the object exists in while imparting a temporal accelleration equivalent to C^5 on the lotus. This hypothetical field would have a rigid dimensional integrity which would allow the mass to go through its many relative dimensional transformations while appearing to remain in a stable three dimensional geometry. Call it a relative dimensional transduction field. Basically it's a dimensional transducer. It would allow you to take a section of the hull of the ship or perhaps the entire hull, and do a spacetime inversion on it. This would have many useful applications, one of which is another one of the bizarre effects I alluded to. You see, since all of its properties are inverted, so are its thermodynamic properties. You could blast though the atmosphere at 10,000 M.P.H. and it would merely cool down the hull. Lasers and particle beams cool it off. It could also bounce cannonballs or artillary shells back at the source with more energy than they started with. A nuclear blast chills the hull close to absolute zero. And so on.

An alternative to this involves a different section of the curve. That would be applying the RDT field to a mass in the C^3 section of the curve. That would probably make it quite a bit more stable, but you'd be losing some of its temporal properties.

Another possibility is one which involves a great deal of precision in the process. That concerns what might be termed 'fractional' time travel. In this case the normal object existing between C^0 & C is dematerialized and transposed with the same object in a past section of the curve which is located between C^3 & C^4. This would effectively 'bump' that object into this section of the curve (C^0 & C), which is our normal spacetime. I think this flip is a little more valid due to the fact that the object in the past has a bit more temporal momentum than the one in the present section and would in all likelihood accept a new position in which it has more energy than it began with. That's one of the principles of temporal energy conservation or entropy. The mass would then be acausal. In that case since you're manipulating the mass directly without changing the physical structure of spacetime, the field will only be needed to initially 'bump' the mass into the acausal state in which it will stay ad infinitum or until the process is reversed.

There are many good reasons for selecting this process. One of them is that you would be gaining quite a bit in the tradeoff, for you now have acausal negative matter with properties that normal negative matter does not have. Negative matter should not be confused with

My Biefield-Brown Anti-Gravity Equation:

$$G = \left[\frac{2.235 * 10^{-13} \left(\frac{(km)A(Eb)}{d} \right) N-1}{m} \right]$$

Where:

G is the induced gravity field in the dielectric (earth=1).
m is the mass of the dielectric in pounds.
k is the dielectric constant of the dielectric (air=1).
A is the area of one plate in square inches.
d is the distance between the plates in inches.
b is the Biefield-Brown constant $2*10^{-5}$.
E is the voltage across the plates.
N is the number of plates

Equation 1

antimatter. They're two different entities. Referring back to the lotus, at positions C^0 to C positive matter is found. (We are composed of positive matter). At positions C^3-C^4 negative matter is found. Between C^4-C^5 negative antimatter would exist, and at C^7-C^8 positive antimatter. But perhaps I should clarify the concept of acausal matter. By "acausal" (or a-causal) I mean that the mass is devoid of the event determinates of cause and effect. It is very close to being in the virtual state, and with the exception of its temporal inversion it would be. Now on to the properties. Any force impressed on the object in the present, or causal side, is actually the effect side of the object due to the inversion. You cannot change an effect. The temporal momentum is infinite. In order to instigate a change in the object the causal side must be changed, and that is in the past. So the mass is also effectively indestructable. The mass of the object could also be engineered to whatever specification needed. To do so would require "tighter" tolerances in dematerialization. If neutron star material were used, the temporally inverted fields inherent in the structure would 'lock out' any electromagnetic or gravitational radiation creating a true gravity shield. The mass could be selectively attenuated by reducing the distance that the original object was transposed upon, thus creating a hull much lighter than any alloy which our present technology can fabricate. Look, ma! no gravity! The shield would also lock out inertial forces so the ship could zip around those right angle turns like the lightcycles in Tron. I have a name for the material. Call it millenium. After all, it can exist for a thousand years without really existing for a second. But I'm digressing.

At C^6 time would move infinitely fast in reverse. The spacelike vectors collapse into a single time vector. At this position space does not exist. Of course it didn't at C^2 either, but in this case time is flowing the other way. You also have temporal simultaneity, the body would be moving so fast that it's everywhen (?) at the same time. Moving at that speed you'd slam into the big bang and careen off into multiple universes (not exactly what I call a fun weekend).

At C7 you would have much the same effects that are seen at the position C with the exception that the time vector is reversed. Think of it as the flip side of time dilation. The faster you travel, the faster your passage back in time will be. If it isn't already evident, this is the section of the lotus that normal tachyons use for propagation.

At C^8 or C^0 the object would be at absolute rest. The time vectors collapse onto the operating space vector, causing an infinite expansion. The body would be everywhere at the same time. Again, this is exactly the same position as C^4 with the space vector reversed. It seems that the ancient theoreticians were right after all, at least with the works that I have encountered. Matter does not exist at absolute rest. The old axiom was that matter was a result of motion, and that in the absence of motion, matter does not exist. Here we have the proof to the pudding. At C^0, without any field propagation, mass does not exist. Another axiom also proves out. That was to the effect that absolute accelleration and absolute rest are exactly the same thing. Once the body accellerates to C^8, it crosses the same point as in C^0, and the same effects are observed there. Of course, these effects are merely relative, and to the object the universe has disappeared instead. This concludes our tour of the velocity gradient. There may be flashbacks from time to time (no pun intended), but those will be only for old times' sake.

Equation 1

My thanks go to Jerry Redfern for helping me out on the more difficult parts of the formula. I must mention first that the Biefield-Brown constant that I used is purely theoretical and probably will change after real world testing has been made. The formula could also be simplified but presently it has been structured for convenience and ease of understanding.

Oh, by the way--there is a formula for a scalar gravity field only, a gravity field which exists without a gradient and just sits there and distorts space. In fact it's merely a variation of the present relation. That model would use a vacuum dielectric and would encounter a problem with high field strengths. That problem would be that high field strengths would make it look as if a massive body existed between the plates, and of course that would throw a spatial vector into the field.

But back to equation 1. G is the induced field strength relative to earth, or 32 feet per second^2. The variable m represents the dielectric mass in pounds. Next is the dielectric constant k of the material selected. A is of course the area in square inches of one plate and d is the distance between the plates in inches. Now comes the iffy part. The Biefield-Brown constant, b, is a hypothetical value. I plugged it into some data and in the instances which I've encountered it has worked. E is the voltage across the plates, which of course must necessarily be D.C. unless you want an alternating gravity field. N represents the number of plates (remember to subtract 1). By the way... if the equation resembles the one for parallel plate capacitance you are absolutely correct. That's what it evolved from.

Quite a few mechanical models have been constructed (and some patented) that would have produced a force vector from a spinning mass. The first few that come to mind are the Dean, Cook and A. C. Nowlin devices. The Dean Drive appears primitive to say the least. The essence of the theory is valid but the model is not very effective. The same could be said for the others. In some of the patents that I've examined mechanical stresses would have been extreme and very close to the limits. One possibility is a variation of a precessional drive; that's a drive based on a series of gyros arranged on the periphery of a rotating disk. Exactly such a model was proposed by Tom Pawlicki in the book "How To Build a Flying Saucer". Now I know that you're probably thinking that the precessional forces would be excessive and the bearing life so short that the device would require constant maintenance. But...if you use magnetofluid or gel bearings that can take abusive forces with a minimum of friction and temperature rise those complications can be avoided. Also due to the losses involved in mechanical models a free energy motor would be mandatory. One theory proposed concerning the Biefield-Brown effect was that it may be caused by electrons precessing in their orbits rather than wholesale orbital distortion that describes a stretched-out ellipse similar to the Nowlin drive. Which brings me to another point. There appears to be an anomaly concerning the multiple armed version of a differential centrifugal force drive in which one end of the orbit has a closer proximity to the shaft than the other. Each pair of weights are mounted on a single shaft which slides through the drive shaft.

Now if the shaft which holds the pair is split or electrically isolated in any way no force results. And if the connection is reestablished the force is as well. It sounds as if an electrical force vector is responsible for the function of this mechanical model. Specifically, it sounds as if a variation of the Biefield-Brown or Searle effect is responsible. I really don't know the exact mechanics involved. I'm just pointing out the errata and hope that at some point in the future the answer will arise.

And now a brief digression back towards the Biefield-Brown effect. After running the program a few times I have come to the conclusion that the possibility of designing a craft utilizing the Biefield-Brown effect is difficult but not hopeless. One possibility is to construct a gravitational traveling wave amplifier with positive feedback. The only way to travel! Of course this makes it a super-regenerative gravitational amp which has some aspects similar to the device that Shinichi Seike outlined as a gravity to electric power generator (you'll see one of his papers in the proceedings of The <u>First International Symposium On Non-Conventional Energy Technology</u>. (see bib.)) Presumably one is the reciprocal of the other. The physical configuration would appear to be a toroid with the plates stacked radially on the exterior of the torus. To be more specific, the plates would run from the outside to the inside of the torus and would give the impression of a series of doughnuts stacked around the large doughnut. There are several advantages to this design, one of which is that the regeneration rate becomes an important factor in gravitational flux generation. The higher the regeneration rate the lower the flux density. If you had a really suped-up drive system, then the regeneration rate would have a fairly high frequency at idle, up in the higher rf range of the electromagnetic spectrum. This would explain the "cooked" soils found at the supposed U.F.O. landing sites. They look as if they've been in a microwave oven too long.

Another point is that if you wanted to generate a scalar gravity field all you'd need to do is construct another toroid on top of the bottom one and run the force vector in the opposite direction. As long as the two cancel out, the scalar value would operate. And that would produce much the same effects that the Searle disk would, namely local spatio-temporal distortion or warpage. That's right--it's a warp drive. Forget about the dilithium crystals.

The warp drive unit and a free energy power supply would be all you ever would need, besides the navigational gear and a few necessary luxuries. The navigational equipment would plot a parabolic vector outside the universe (or to the single-universe chauvinists, a vector outside 3 space).

A variation of this concept would be a warp transposer. Quite literally it would create a scalar wormhole, the end of which would project outside tri-dimensional space and loop back along a parabolic trajectory to the destination point. This unit would contain the warp in a bottle-like arrangement without spillover into the surrounding space. Since you could plot courses along the lotus the transposer could operate through space <u>and</u> time as well. You could 'beam' from New York to Los Angeles and arrive five minutes before you left. Or send packages the same way. You truly could get it there yesterday. To send something you'd punch in the space-time coordinates and the computer would calculate the trajectory. Then the object would be placed on an alignment grid and the field switched on. The mass would then accellerate along the field vector. When the temporal momentum of the mass exceeded the dimensional change per unit time of the field, it would emerge at its destination. Of course the opposite would be employed to get the object back onto the transposer grid. In that case the field would be projected or transposed onto the object to be "snatched". Then the object would exist in both the transposed and the original spatio-temporal field. Remember that the field created by tangible matter is between C^0 and C on the lotus. The whole objective is to dump energy from the field into the mass to be transported. Once this is accomplished the mass accellerates along the field vector. It "falls" outside space-time. The bottom of the well would be the transposer grid. Since large amounts of temporal momentum would be dumped into the mass over large distances, considerable damping or compensation would be required which would place a limit on the maximum distance that the mass could be transported safely. In other words, if you tried to "beam" in from a large distance you'd most likely arrive with a bang.

Tunguska, Siberia Blast in 1908

Which leads us into another subject somewhat similar. What really happened to create the tremendous explosion which leveled a large area of Siberia 1n 1908? A quasi-accepted theory involves a large meteor impact. This seems implausible, however, considering the lack of an impact crater anywhere near the site. Another theory involves a U.F.O., and that particular view holds that a spacecraft with a damaged power generation system blew up due to a runaway chain reaction on board. In light of the new insight toward gravity, another theory suggests itself. Incidentally, I'm going to have to invoke that mysterious beastie: hyperspace.

Let's assume that black holes do in fact exist, as well as white holes. These bodies seem as severe a model of gravitational wave theory as you could get, but there is another type of cosmic pac-man which is even more insidious. That type would involve a black hole without a vector to its field. Think of it as a black hole and a white hole of similar masses superimposed over each other. Or of a black hole of negative matter superimposed over a black hole of positive matter. One would have a "push" to its field, and the other would have a "pull". Now here's the interesting point: when the two fields superimpose, the force vectors present in three dimensional space will cancel out. But there's still another force vector to deal with, and that is one that <u>does not</u> go away. That force vector is in four dimesional space and warps, or distorts it. What this means is that the volume of space associated with it appears to be a disembodied space/time warp. Of course there is a mass there; in fact quite possibly two. It's just that it's undetectable as far as its gravitational influence is concerned. This body would be gravitationally neutral so nothing attracts it and nothing is attracted into it.

There could be three ways to detect it. One way would involve using a kind of cosmic plumb line. If you drop a mass into it and it undergoes a violent dimensional change, then you've found it. The same method could be used to "map out" the area to find the borders. Another method could rely on the fact that this body has what appears to be a vast amount of space on the inside of it. Since light would still obey the inverse-square law once it passed the event horizon, a great deal of attenuation would occcur as seen from an outside observer. In fact, the body would have the same effect on light as a concave lens. For example, if light had ten miles to travel on the outside, and ten million on the inside, the light exiting the horizon on the other side would seem to have traveled that much farther. Another method would involve measuring the time delay between two signals; one sent from the source to the observer on the outside of the event horizon, and the other sent from the source and through the zone of space under the influence of the body. The delay would be proportional to the amount of space on the inside versus the outside.

How large could these beasties get? The size would be related to the superimposed masses. Conceivably, the smallest they could get would be in the neighborhood of 10^{-33} cm. Which would also be the size of a theoretical particle of free space, called the chronon. Could there be a connection? Quite possibly so. The thought occurs to me that if you ran all four quadrants of the lotus into each other you'd get much the same properties that the chronon has. In that case, the field density in the inside would be so high that any mass would appear to be geometric points from an outside point of view. In the inside, however, an oberver would see an infinite volume of space and three-dimensional, tangible matter. So there would be worlds within worlds. (Please bear with me. All of this preliminary stuff is necessary to understand what I've got that follows). If the superimposed masses had a substantial rotational speed, something else happens. A hyperspatial skew would occur which would produce a wormhole in space. This would be related to the gravitational coriolis effect generated by the bodies. Actually, a gravitational vortex is the easiest way to understand it. A gravitational tornado or whirlpool would be generated which would induce an accelleration outside spacetime. You see, since the masses cancel out any attraction that the field could have, the only acelleration that would occur is actually outside spacetime itself. So the mass acellerates. To where? To the business end of the vortex which is transposed upon normal space. And all the way it's gaining kinetic and temporal momentum at the expense of the body which is propelling it. Let's call the superimposed mass combination a C.V.E. That stands for charged vacuum enviodment. After all, all that's left between the gravitational tug-of-war of the two bodies is a charged vacuum which is apparently empty. Another point I'd like to clarify is that in this scenario, no such thing as a white hole would be observed at the business end unless the C.V.E. happened to drift into a matter-rich area of space (like the core of a planet, for example). Even then, no superheating of otherwise cold matter would occur unless the C.V.E. is at a distant location. In that case you'd have what is termed a "cosmic gusher". The distance between the C.V.E. and its business end would depend on its mass.

So much for the preliminaries. What would happen if a chunk of rock just happened to drift into a rotating C.V.E., one in which the two masses counterrotate in relation to each other? It would be acellerated through hyperspace and gain energy at the expense of the angular momentum of the two superimposed masses. And if the C.V.E. were very massive, it would gain a phenomenal amount of energy by the time it reached the exit point and literally slam into its destination, releasing all of its energy at once. I have no idea how often many of these galactic slam dunks occur. But with a massive C.V.E. thousands or even millions of light years away, who knows? I think this may have happened in 1908, in Siberia. Perhaps a wandering asteroid found a C.V.E. and suddnly found itself in Siberia. When it "popped" into its destination, it might become superimposed on top of the gas molecules which just happened to occupy the same space. Which might explain the fission fragments found near the impact(?) site. After all, what would happen when, say, an iron nucleus slammed into an oxygen or nitrogen nucleus? An immediate

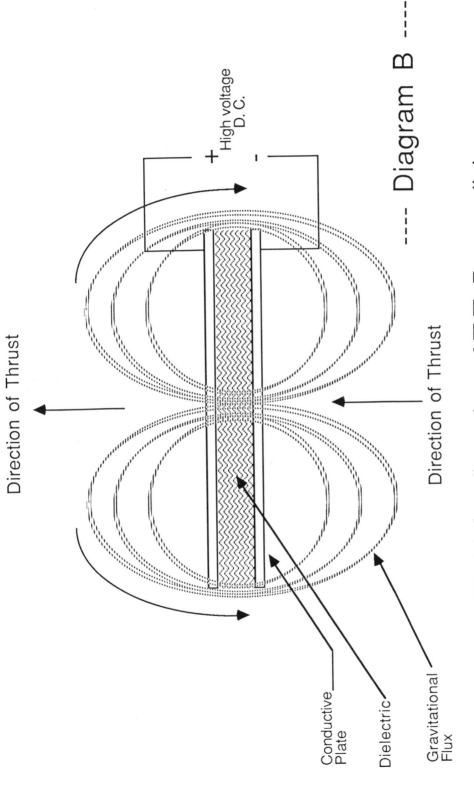

Field Configuraion of T.T. Brown disk according to Ether-Technology --- Diagram B ---

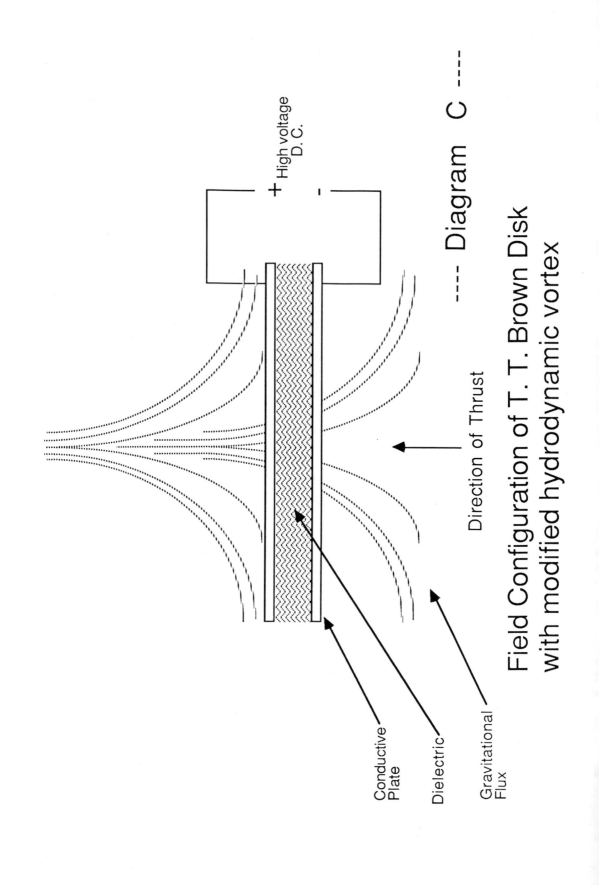

fission/fusion reaction would ensue. That additional energy would probably be insignificant compared to the enormous amount of kinetic energy contained by the asteroid. Of course the same thing could happen to a spacecraft, if that theory is valid. And so... Kaboom!

Now on to an explanation of the remaining diagrams.

Diagram B

Diagram B shows the gravitational flux flow around the Biefield-Brown Disk according to Rho Sigma in Ether Technology (see bib.). The direction of thrust is actually the direction that a mass would take as it "fell" into the field. Usually the impression that one gets from the term "thrust" is a push rather than a pull, so that term is somewhat misleading.

Diagram C

Diagram C shows a Biefield-Brown Disk with what I term a modified hydrodynamic vortex. This flux flow is markedly different from the one in diagram B. This configuration is also much more prone to "matter snatch" than the one in diagram B. It would seem that in diagram B the field produces more of a pinching action than the one in C. The matter snatch effects would be interesting. The one in B would produce a kind of lateral compression of the soil with a small cross section pulled up where the force vector is strongest. The one in C would cut round, neat holes in the soil. The soil would literally stick to the underside of the disk.

Diagram D

Diagram D shows the gravitational field configuration of a T.T. Brown disk with the conventional radiation pattern. This view shows the field radiating away as an electomagnetic field of radiation normally would. This view (diagram D) would only be correct with a disk composed of a superdense mass, such as that suggested by Robert L. Forward (see Bib.). It would not be correct with the observed effects linked to the artificial generation of a gravity field using an electric field.

Diagram G

Diagram G shows the gravitational field configuration of the Searle disk and the resultant electric field potential across the face of the disk. This shows the similarity between a Searle disk and the operation of a homopolar generator, or "N" machine, as suggested by the work of Bruce Depalma (see bib.).

Diagram I

Diagram I shows my version of a Schauberger device using four torqued fluid vortices. In theory, the operation of torqing a fluid vortice would be identical to torqing a gyro, since the gyroscopic moment of inertia is similar. All that would be needed to test this would be a pump motor of sufficient capacity, fluid baffles, and the associated piping. In this device, the curvature of the pipe in which the baffles are placed and the placement of the fluid baffles would be somewhat critical. Think of it as a liquid "Laithwaite engine".

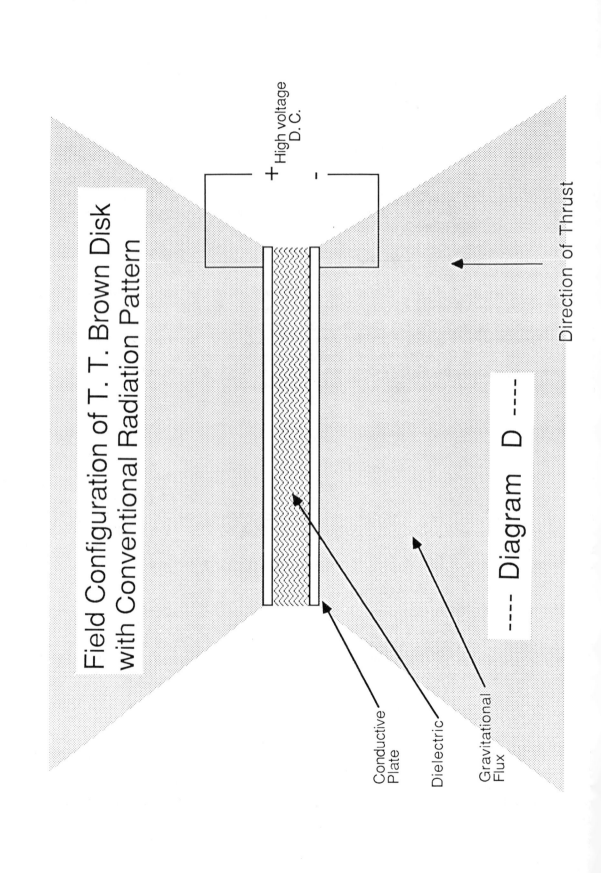

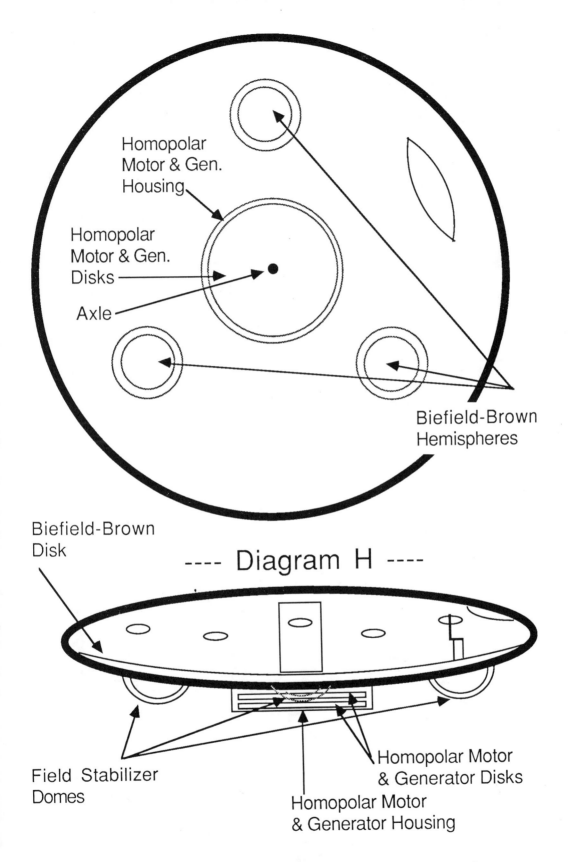

---- Diagram H ----

Antigravity Craft With Biefield-Brown Effect Drive

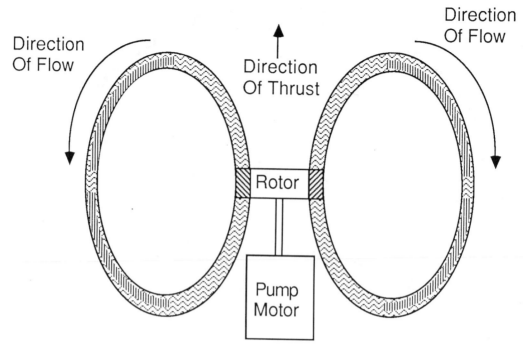

---- Diagram I ----

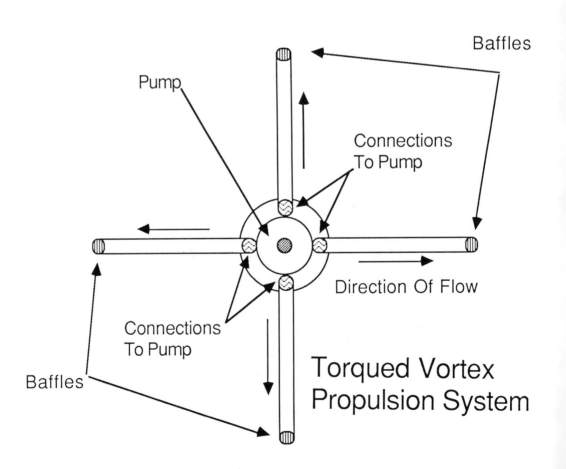

Torqued Vortex Propulsion System

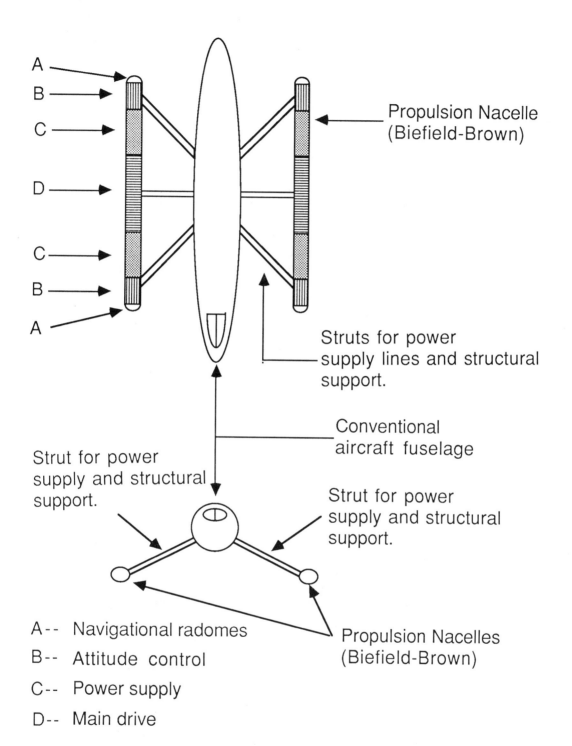

**Conventional Fuselage Configuration
With Gravity Drive Retrofit**

---- Diagram J ----

Diagram H

Diagram H is my idea of what a ship would look like using the best of the alternative technologies available. This version is in the shape of a discoid craft, not really a true saucer. The homopolar motor/generator resembles the N-1 Power Generation System of Bruce DePalma (see Bib.). This version has a few modifications on it, however. One of which is a tremendous increase in size, and another would be a specialized power supply which would feed into the main drive. One addition to this design which would be absolutely mandatory is the field stabilizer domes. When the main drive is in operation and a fairly high field amplitude is reached, a kind of a "gravitational coriolis effect" would come into operation and would tend to torque the ship and make it spin around like a frisbee. This prevents that undesirable operation by generating "mini-vortices" that destroy the angular momentum in the gravitational field thrown out by the main drive. Top speed? The top speed would be proportional to the power output of the main drive and the capacitive load of the Biefield-Brown drive. With state-of-the-art materials and technology this craft could possibly do fifty percent of the speed of light. This version would provide a pretty conservative accellaration as far as UFO's go. Some UFO's accellerate as fast as 32000 feet per second^2. This one would be a model T, more or less. Its accellaration would be approximately 960 feet per second^2. To get some idea of the proportions of the craft and accommodations, the diameter is thirty feet across. With all the space that would be taken up by navigation and life-support equipment, the only space that remains would be enough for three or four passengers and one pilot.

Diagram J

This diagram shows a more conventional design than the one outlined above, in fact it can even be a retrofit to a Lear fuselage. But there is a tradeoff. The field generated by the nacelles is completely self-contained and doesn't protect the crew from the effects of accelleration. So this is primarily a shuttlecraft, and of course with the limited accelleration, comes a limit to speed and range. In other words, this is not an interstellar craft. The top speed (if the fuselage were retrofitted to operate in a hard vacuum) would probably be approximately 1000 miles per second. The range would be limited by the amount of supplies the ship could carry. The two "caps" on either end of the nacelles (A on the diagram) are the navigational radomes used during flight to guage distance to the destination and detect foreign objects which might damage the craft. The items designated as B are the Biefield-Brown propulsion systems used for attitude control. These are electrically isolated from the main drive and have considerably less power. The items designated as C on the diagram supply power to the nacelles. Two are supplied per nacelle with one used as a backup unit.

This would probably be either a stationary armature generator type or an N-1 power generation system. Struts run from the nacelles to the fuselage providing structural integrity and power to the fuselage. The main drive is a gravitiopile comprised of Biefield-Brown disks stacked vertically to provide lateral accelleration. The number of passengers would be limited to the fuselage selected.

Diagram K

Diagram K shows the results of a great deal of research into the relationship between the gravitational field and the other forces in nature. It also shows what I term the "Energy Gradient". This diagram shows the relationship between two primary fields named Phi and Psi whose phase relationships combine to form all the physical forces in nature. I'll start with the bottom of the diagram, the one marked "Free Space". In this position, Phi and Psi are 180° out of phase in what is termed as destructive interference. Since the phase difference between the two fields defines the physical dimensionality of free space, in this position no space exists. This diagram evolved out of the familiar electric-magnetic-gravitic triangle that is shown in diagram M. The earlier version did not explain the propagation characteristics of an electromagnetic wave, which made it somewhat

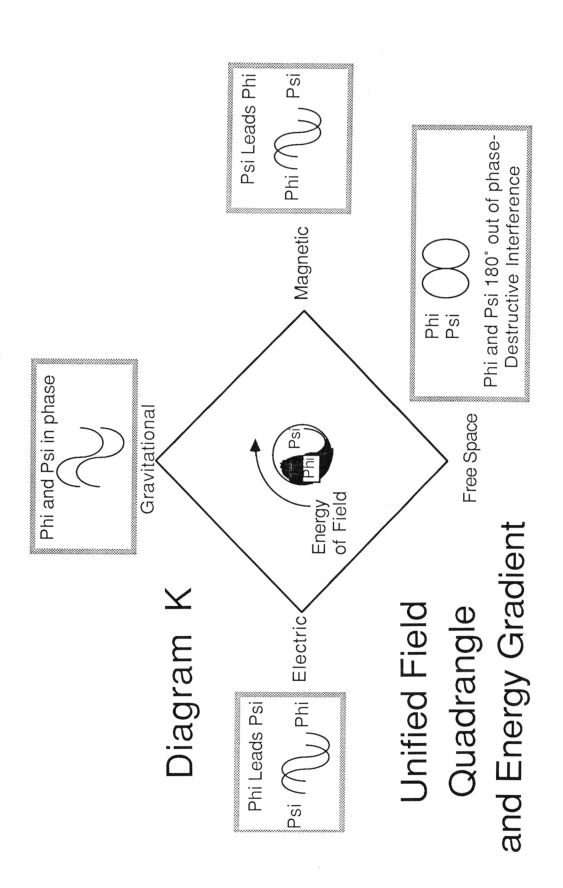

incomplete. The earlier theory was very mechanistic and ignored the dynamic capacity of the space in which the wave propagated. Hopefully, this will shed new light on the puzzling phenomena which is observed in the universe in which we live.

In diagram K I have pair-coupled two fields which are named psi and phi. These two fields are never seen alone. Each is always pair-coupled to the other. The phase angle between the two, depending on which one is leading and which is lagging, determines which force exists in free space.

Beginning at the bottom of the diagram, I have shown that the two fields in destructive interference resemble free space. They would but... Free space in that diagram would have a zero energy density and thus would have no dimensional charateristics. Four dimensions, length, breadth, width, and duration would collapse into one. In the space that we are familiar with, there is always some amount of angular dispacement between the two. Einstein had an adage which he used to explain his theory and which I'll repeat since it has some measure of validity here. Mass tells space how to curve and the space tells the mass where to go. But gravity (or the curve) also tells space how to expand. A gravity field induces an angular displacement away from 180°. That's one way of looking at it. From another perspective, the angular dispacement induces compression of the standing wave "bubble" of a particle and thus by shrinking it down gives the illusion that there is more space around it. Either way you'd get the same effect.

Moving along the diagram in a clockwise direction, we see the electric field. In this position phi leads psi by 90°. It would have an intimate connection to free space and to the gravitic field. In theory then, it should be possible to affect both by using the electric field. It is my contention that the Biefield-Brown effect, the Searle Effect, and other mechanisms used to generate electro-gravitic fields does exactly this. A good rule of thumb to use with the diagram is this: Each field affects the fields that are at right angles to it in the diagram. All other fields that are not shown in the diagram exists between any two of the fields on the diagram.

Now on to the gravity field. In this case the components phi and psi are in constructive interference with each other. A very weak gravity field would have some of the properties of a strong electric field, and vice-versa. With a 0° phase angle between them, the phi and psi fields would produce an infinite volume of space. This would be diametrically opposite to the position below it which is shown as free space. Of course from the other perspective, the abscence of angular dispacement induces infinite compression of the standing wave "bubble" of a particle and thus by shrinking it down to a geometric point gives the illusion that there is nothing there. From the position on the diagram another inference may be made. It would be imposible then to have a strong gravity field that did not spill over into the magnetic or electric sides. And if that spillage pair-coupled, there would be a direct conversion of gravitational to electromagnetic energy. That possibility may be valid but I think another is far more likely. If a portion of this "leakage" went into self-oscillation an electromagnetic wave would be created. This very mechanism could cause heating of massive bodies of planetary proportions. It could quite possibly attain as much as 10^6 tons per second in a body approximately 330,000 times as massive as the earth. This conversion is accomplished at the expense of the mass, however, and the body in turn suffers from a form of cosmic dry rot. Conceivably, this could be the cause of the 3° K. backround radiation in the universe. More will be discussed in my upcoming book.

Proceeding now to the magnetic side we see that psi is leading phi by 90°. If the magnetic field went into self-oscillation it theoretically would fluctuate between a gravitational field and free space with zero energy (also zero volume). Needless to say, this would make for some very interesting effects. Magnetic material saturated with this self-oscillating field would show relativistic effects standing still. Supersaturation could even possibly induce dematerialization.

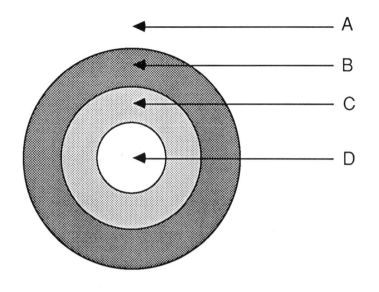

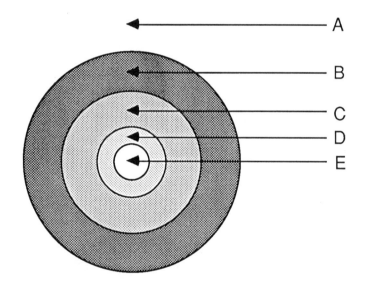

A: Free Space
B: Magnetic Field
C: Gravitational Field
D: Electric Field
E: Free Space (Zero Energy)

Cross Section of Magnetic Field Line

---- Diagram L ----

Diagram L

A magnetic field line is a special case unto itself. At the operational level of the field line, which is in diagram L, we see that at the core of the field line in number 2 a singularity exists. At the "surface" of the field line, psi is thrown 90° forward. Penetrating the surface of the field line we come to position C on the diagram, which is a gravitational field in which phi and psi are at a 0° angular displacement from each other. This gives an infinite value to the gravitational field, and thus it exhibits the same properties as a singularity. It also produces an entity which has (for all practical purposes) a dimensionally transcendent geometry within itself. Moving on to position D, we see that an electric field is infolded within the magnetic field line. This means that if it were possible to rip a magnetic field line apart, or even flip one inside out, an electric field would be observed. In position E, a zone of Free space exists at zero energy. This of course means that at position "E" a condition exists that gives the zone a finite diameter on the outside and zero volume on the inside. This also implies that if the technology permits to inject a controlled source of energy into one end of a magnetic field line and into zone E, then that energy will "pop" out the end of the field line at the same instant that the signal was injected at the entry point. That signal would propagate at a seemingly infinite speed because physical space does not exist in the center of the magnetic field line in this scenario. And if the signal were injected and allowed to loop through a field line, then you would have a signal moving backward in time. That signal would probably move back to the point prior to the existence of the field line and then would exit and dissipate its energy. This could be a pretty simple way to induce temporal feedback, if it could be realized.

Table A

This table contains various effects that would be observed for a physical mass progressing through the velocity gradient. This table plots relativity and beyond. beginning with position C, the table notes a positive value for T, which is time. A positive value is noted for S, which denotes space. These two values mean that at position C the object exists in positive space and positive time. Value M denotes the momentum of the mass. In this case, $Aleph^0+$ means that the body contains an infinite amount of positive momentum. The aleph series was created for counting quantities greater than infinity. $Aleph^0$ is infinite. $Aleph^1$ is infinity squared. $Aleph^2$ is infinity to the third power, and so on. Simply put, it merely counts powers of infinity. Under distortion effects, time dilation is noted. This means that to an outside observer, the subject is moving more and more slowly as time slows down for him. Also noted is spatial contraction, which means that to the subject, much less space seems to exist between the subject and the destination point. The next point is somewhat debateable. This is seen on the tabe as "Local Spatial Expansion", and is based on a rather strict interpretation of Einsteinian space. In this case, the subject and the ship are contracting (actually physically shrinking) in the direction that the ship is moving. So the ship and everyone on board becomes thinner with an increase in velocity. There is another effect that would be seen with dimensional transduction fields. That effect would still cause a spatial contraction in the direction of movement, but if the aceleration is outside 3 space, then the atomic particles that comprise the ship and its' occupants contract in three dimensions until they disappear into a collection of geometric points. The next point on the table is position C^2. In this position the body is moving at the velocity of the speed of light squared. The body exists in positive time as seen by the + under "Value T". Value S has "Null", which means that the object does not occupy physical space. Value M is also "Null" which means that the body also has no momentum. Under "Distortion Effects", positive contraction of time is noted as the body moves infinitely fast into the future. Spatial Suspension denotes that from the perspective of the occupants of the ship, our universe ceases to exist. Temporal simultaneity (with the + symbol attached to it) is another assumption which may or may not be correct.

Position	Value T	Value S	Value M	Distortion Effects:
c	+	+	$Aleph^0$ +	Time Dilation Spatial Contraction Local Spatial Expansion
c^2	+	Null	Null	+ Time Contraction Spatial Suspension +Temporal Simultaneity
c^3	+	−	$Aleph^1$ 0	Negative values for physical properties of free space.
c^4	Null	−	Null	Temporal Suspension − Spatial Simultaneity
c^5	−	−	$Aleph^2$ −	Space/Time Inversion
c^6	−	Null	Null	− Time Contraction Spatial Suspension −Temporal Simultaneity
c^7	−	+	$Aleph^3$ −	− Time Dilation Spatial Contraction Local Spatial Expansion
$c^8 - c^0$	Null	+		Null Temporal Suspension + Spatial Simultaneity

---- **Table A** ----

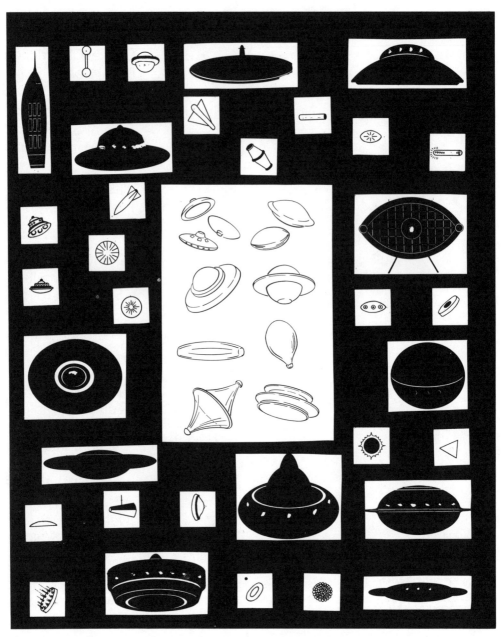

What UFO shapes were people seeing in the skies? All kinds: balls, cigars, doughnuts, disks, cylinders. At center of this montage are the ones most commonly reported. Some observers have reported UFOs growing larger or smaller while still in view; others that they glow, pulsate, or change color. Some are heard to hum; others are silent. Nearly all agree they can perform fantastic maneuvers—zig-zagging, turning at 90-degree angles, hovering, or reversing course at unbelievable speeds.

GREAT RESEARCHERS INTO GRAVITY NO. 1

ALBERT EINSTEIN 1879-1955

Quote: "Gravity cannot be held responsible for people falling in love."

Einstein's work and theories on the "unified field" was a landmark in gravity control.

In a 1944 letter to Truman Einstein suggested the creation of an American-Russian state that would control the world.

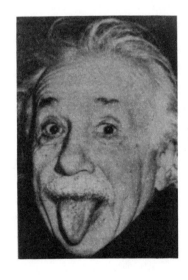

While walking about in one of his typical dazes, Einstein managed to walk directly into an open manhole. A photographer who captured the event was bribed into supressing the picture.

As a youth, Einstein was so inept at academics and sports that his father believed him retarded, even as an adult. Einstein could not even master the complex task of driving a car.

Einstein married his divorced cousin, Elsa Einstein. In a statement about his deep love for her, Einstein said, "Mrs. Einstein is an excellent cook. If she weren't, I would divorce her."

Einstein once said, "The hardest thing in the world to understand is income tax."

While Einstein was residing in Princeton, the elderly scientist would often forget where he lived. Once stopping a stranger on the street, he exclaimed, "Hello, I'm Albert Einstein. Could you tell me where I live?"

Einstein was offered the chance to become President of Israel in 1952, but turned it down saying, "I know a little about nature, and hardly anything about men."

It may not be an unattainable hope that some day a clearer knowledge of he processes of gravitation may be reached; and the extreme generality and detachment of the relativity theory may be illuminated by the particular study of a precise mechanism.

—Albert Einstein

It would of course be a great step forward if we succeeded in combining the gravitational field and the electro-magnetic field into a single structure. Only so could the era in theoretical physics inaugurated by Faraday and Clark Maxwell be brought to a satisfactory close.

— Albert Einstien, "Mein Weltbild"

ELECTRO-MAGNETIC PULSE
by Karl Kruszelnicki

One single nuclear weapon exploded high above Alice Springs would wipe out all the solid state electronics in Pine Gap and most of the telephone communications and electrical power cables in all of Australia. Also it would probably electrocute people in Sydney, if they were unlucky enough to be leaning against a long metal fence, or having a bath if the bath was filled with water from metal pipes.

The thing that can electrocute you from such a great distance is called EMP or electromagnetic pulse. This happens when a nuclear weapon is exploded at a distance of 40 to 400km above ground - that is, not *in* the atmosphere, but above it. When the weapon explodes vast numbers of gamma and X-rays are unleashed, which cover large distances before they start hitting air molecules. Electrons are then stripped off from the air molecules, and the electrons start to twirl down and around following the lines of force of the Earth's magnetic field. As they spin they start acting like tiny radio transmitters, but they are putting out huge amounts of radio energy. EMP is very destructive because any length of metal will pick up this radio energy. Look at the cars running around Sydney with coat hanger aerials as proof of this. So if you have a telephone line or a power line which crosses a continent, say the continent of Australia, or Russia, China or Europe, it will generate something like 10 million volts and 10,000 amperes. This is enough to burn through any insulation we've got today. If you touch a telephone line or a radio dial when the nuclear weapon goes off you're in big trouble.

EMP was first noticed by US scientists back in 1962, when they had a little nuclear experiment called "Starfish Prime". They exploded a one and a half megatonne nuclear weapon 400 km above the ground level of Johnston Island in the Pacific. 1500 km away in Hawaii there was massive electronic destruction as three hundred street lights blew up, burglar alarms triggered off, power lines fused and TV sets exploded. The EMP will burn out transistors in integrated circuits, it will burn out radios and it will burn out TVs.

Radio valves are 1 billion times more resistant to EMP than integrated circuits. The Russians know about this, and in their MIG 25

Foxbat interceptor fighter they use, not only valves to stop EMP, but also a double skin on the aircraft to stop electromagnetic radiation penetrating. United States investigators found this in 1976 when a Soviet pilot defected to Japan and they pulled the plane to pieces. They started laughing because they thought, "Valves in 1976, how primitive!" But late in 1977 the Pentagon rewrote the handbook on the effects of nuclear weapons and said "Yes, go ahead and use valves, they are a better device". In fact, the Soviets are very up to date with EMP. As one recent Soviet war manual said, "To achieve surprise in a war, high altitude nuclear explosions can be carried out to destroy systems of control and communication". Any EMP would destroy the electronics of satellites whether they are spy satellites or communication satellites.

Can you protect yourself against EMP? Well it's not impossible but it is very difficult. You can't harden a telephone line, forget it, but you can harden a single device. You could wrap a radio in a whole lot of alfoil. If you were flying in a modern aeroplane and a nuclear weapon went off a few thousand kilometres away, the plane would fall out of the sky like a bunch of car keys. In 1970, Boeing started trying to harden some 747s and they wrapped the cables in lead and put wire mesh on the windows. When they tested it they found that 12,000 circuits, essential for the running of the aircraft, had fused. Later, they started from scratch, and hardened the 747 right from the very beginning. It didn't have any windows and it cost five times as much as a normal 747, but it was hardened - there was only one ever built. In 1980 the Americans built a device called Trestle to generate and test the effects of EMP. It's called Trestle because it looks like a giant railroad trestle bridge twelve storeys high and made entirely of wood - Douglas Fir in fact. There is absolutely not one scrap of metal in it. It cost 60 million dollars and it can hold a fully loaded B52. But the problem is that it can generate only about half of the EMP that a nuclear weapon would produce. In 1981 Reagan put aside 20 billion dollars, about one fifth of Australia's gross national product, to strengthen and rebuild the communications system. In Switzerland they are serious about EMP - all of their military computers are buried 600 metres underground in the Alps.

Now we have the situation where one single nuclear weapon exploded high above the atmosphere could wipe out all the electronic communications and power lines over an entire continent and so cripple an entire nation.

There are thousands of satellites up there and any one of them could be loaded with a nuclear weapon. Right now somebody could be pressing the button.

So the meek, that is things made of valves and wood, will inherit the earth. Don't bath, and you might make it too.

Karl Kruszelnicki was born in Sweden and spent his first year in Australia in a refugee camp. He has a Bachelor's degree in Mathematics and Physics, and a Master's degree in Bio-medical Engineering. Karl has worked as a labourer, physicist, tutor, researcher, roadie for rock and roll bands, cab driver, filmmaker, car mechanic, scientific officer, bio-medical engineer and radio journalist; he is currently studying medicine at Sydney University.

© Text - Karl Kruszelnicki

All rights reserved.

First published in 1984 Boobook Publications Pty.Ltd., PO Box 238, Balgowlah, NSW 2093, Australia.

Ref. **Omega**, *May/June 1982, pp 6,8,18*
  Science 83, *February, pp 40 - 49*
  Electronics Australia, *April 1983, pp 14 - 19*
  Scientific American, *January 1984, pp 23 - 33*

HISTORY OF THE XM-5 COMMUTER

The development of a VTOL aircraft for use as a mass transit vehicle has been anticipated since the end of World War II. No aircraft to date has proven itself to be a realistic candidate largely because of cost and operating difficulty of existing VTOL aircraft (mainly helicopters).

A practical VTOL aircraft, to gain wide acceptance, must be simple and safe to operate, inexpensive to manufacture, and economical to maintain.

After two decades of intense theoretical and experimental studies, Moller Corporation has created a new class of VTOL aircraft that meets these requirements. A two passenger version, designated the XM-4, is presently in test. It is powered by eight air-cooled rotary engines and composed of a two-piece fibreglas airframe. The airframe acts as a lifting body while the thrust modules provide lift in hover and thrust in forward flight.

---

1. Moller, P. S. "A Radial Diffuser Using Incompressible Flow Between Narrowly Spaced Discs" Transactions ASME Journal of Basic Engineering. March 1966 pp. 155-162.

2. Moller, P. S. "The XM-5 COMMUTER — A History And Analysis of VTOL Aircraft" Selected Distribution Internal Report of Moller Corporation. March 1982

3. Moller, P. S. "A New Concept in Airborne Vehicles" International Congress of Agricultural Aviation at Queen's University, Kingston, Ontario. August 25, 1969.

XM-4 in flight test.

A five year test program to refine the external aerodynamics began in 1976. This study used one-sixth scale radio controlled models that were extensively wind-tunnel and flight tested. The goal and result of this program was to secure a low drag coefficient and a highly stable aerodynamic configuration for this class of aircraft.

Model test configuration.

The XM-5 COMMUTER is a comparatively inexpensive four-passenger VTOL aircraft which evolved from these model and XM-4 tests. It uses the latest aerodynamically determined shape together with the propulsive and hover stabilizing technology of the XM-4.

The recent production of suitable piston and rotary engines and the development of a low-cost rate sensor has made the XM-5 COMMUTER an immediately viable vehicle. In addition, the F.A.A.'s willingness to consider user assembly time in place of fabrication time makes the XM-5 COMMUTER immediately marketable under the experimental aircraft category.

We have, therefore, given top priority to the development, manufacturing, and marketing of this four-passenger model. We expect production to begin within 12 months.

ADVANTAGES IN USE

Because of its protected propulsion system, VTOL capability and small size, the XM-5 COMMUTER can operate in very confined areas. Since it can also take off from water, landing sites are unlimited.

The XM-5 COMMUTER will provide low door-to-door transport times with a cruise speed three times that of the automobile and twice that of the light helicopter. By achieving a low profile drag coefficient, it is particularly efficient at higher speeds where induced drag is small. The following figure compares the performance of the XM-5, helicopters, and light planes. The comparison criteria, PAYLOAD × VELOCITY ÷ POWER, is a measure of the transport efficiency.

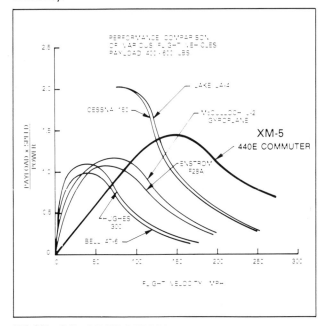

EASE OF OPERATION

The XM-5 COMMUTER uses eight individual powerplants which operate collectively for manual height control, but separately and automatically for pitch and roll rate control. Each engine has its own assigned single axis fluidic-electronic rate sensor which monitors motion about a prescribed axis. The thrust level of the appropriate engines are then automatically adjusted to provide the pilot's desired rate of roll or pitch independent of random aerodynamic inputs. Thus, the most difficult aspect of VTOL flight is eliminated by a highly redundant automatic stabilization system. The XM-5's ease of operation is also enhanced by its inherent insensitivity to the all-important wind shear or gust. This results from a favorable interaction between its wing loading, wing lift-slope, and lift fans.

The XM-5 COMMUTER can use automotive gasoline and therefore need not access conventional airports.

OPERATIONAL COSTS 440E XM-5

The use of multiple lifting fans, variable camber duct exit vanes, and differential thrust control eliminate the cyclic and collective blade pitch control required in a helicopter. The directly driven fans eliminate both transmissions and clutches while the after-swirl vanes eliminate the need for an anti-torque tail rotor. These components account for the helicopter's complexity and maintenance costs.

The average airplane suffers from the indirect costs of under-utilization. The 440E's versatility will result in a cost/mile well below that for today's light plane if it is used in a practical everyday commuter role.

PILOT REQUIREMENTS

The 440E is designed to fly as a conventional fixed wing aircraft except during take-off and landing where the ducted fan airflow will be directed downward for lift. The minimum licensing requirement is expected to be a student pilot's license. When F.A.A. certification of the 440E is achieved, the required pilot licensing should be well established.

NOISE 440E XM-5

The 440E's unique ability to operate near or within highly populated areas requires an emphasis on noise suppression. The ducted fans are designed with a moderate thrust loading and low tip speed in order to generate minimum noise. Hover tests have demonstrated a perceived noise level of 85 PNdb at 50 feet. This is less than 40% of the noise level produced by a Cessna 150 during take-off, and comparable to that for rush hour city traffic.

SAFETY

Any widely used VTOL aircraft must deal easily and safely with powerplant failure. The 440E's multiple engine design provides a particularly high reliability. For example, if a single engine has a reliability of .99, then the reliability of six out of eight engines is .99995.

A 6:1 glide ratio allows conventional power-off landings to be undertaken should complete powerplant failure occur. It is also possible to maintain forward flight with only two of the eight engines operating.

PILOT REQUIREMENTS

The XM-5 is designed to fly as a conventional fixed wing aircraft except during take-off and landing where it would operate as a ground effect machine. Historically, ground effect is assumed to extend to a height equal to the wing span of the vehicle or to rotor diameter in the case of a helicopter. By this definition the XM-5 could operate up to 15 ft. from the ground without requiring a pilot's license. When operating above 15 ft. the minimum requirement would be a student pilot's license.
We are presently working with the F.A.A. in order to establish operator licensing requirements consistent with the demonstrable safety of this aircraft.

PRODUCTION SCHEDULE AND COST COMPARISON

The construction of the necessary tooling for production of the XM-5 has been initiated. The first production aircraft will be complete by July '83 with delivery occurring before the end of 1983. The level of production is yet to be established and depends on orders placed before the end of 1982. The two major factors determining the XM-5 selling price are its projected production level and ability to be sold under the experimental aircraft category. The experimental category does not require F.A.A. approval of the aircraft, but rather requires user participation in the construction of the finished vehicle. In the case of the XM-5, where factory fabrication time is low (airframe fabrication requires 16 factory man-hours) the user's participation becomes one of assembly alone. Based on our expected fabrication time, it is anticipated that approximately 125 user hours will be needed to meet the F.A.A. requirements within this experimental category. Under these conditions, the XM-5 COMMUTER would sell for approximately $20,000 at the 10,000 unit/year level.

PRODUCTION FACILITIES

Moller Corporation which was formed in 1968 recently expanded its research and production operations to the Davis Research Park. Much of this new 34,000 sq. ft. facility will continue to be used for manufacturing our SuperTrapp high performance silencing systems.

Approximately 8,000 sq. ft. of this plant is available for XM-5 related development and manufacture. This is adequate space to assemble computer aided controls and thrust modules for approximately 1000 aircraft/year. An additional 35,000 sq. ft. would need to be added in order to reach sufficient component production for 10,000 aircraft/year. The fibreglas airframe will be manufactured at separate facilities.

XM-5 COMMUTER

Specifications

Passengers	4	Hover Ceiling	9,500 ft.
Cruise Speed	165 MPH @ 55% power	Operational Ceiling	16,000 ft.
Top Speed	215 MPH	Fuel Capacity	42 gallons
Rate of Climb	3,060 FPM @ 75% power	Empty Weight	640 lbs.
Max. Range	320 miles, 20 min. reserve	Gross Weight	1,420 lbs.
Max. Payload	600 lbs.	Powerplant	Multi-rotary (8 x 40 HP)
Design Payload	475 lbs.	Dimensions	15.2'L, 14.8'W, 4.6'H

Moller Corporation is financially able to produce the XM-5 in limited quantities through funding from its present operations. However, the market response to the XM-5 COMMUTER suggests that even a 10,000 unit/year level might be insufficient to meet either the civilian or military interests. We therefore anticipate that a portion of XM-5 related growth capital will continue to be met by new or existing investors. We are in the process of forming a separate division of Moller Corporation to manage this aircraft operation.

The XM-7 is a single place VTOL aircraft that is extremely compact. It has the potential of being classified as an ultra-light aircraft subject to the F.A.A.'s interpretation of its minimum stall speed. The XM-7 will use three ducted fans identical to those in the XM-5 and be equipped with a hydrogen peroxide lift back-up system to provide safe operation during the VTOL mode should a powerplant fail. The selling price of the XM-7 is expected to be approximately one-half that of the XM-5.

XM-7 Specifications (Projected)

Passengers	1
Cruise Speed	85 MPH @ 65% power
Top Speed	105 MPH
Rate of Climb	1,700 fpm @ 75% power
Maximum Range	115 miles, 15 min. reserve
Payload	175 lbs.
Hover Ceiling	6,500 ft.
Operational Ceiling	12,500 ft.
Fuel Capacity	11 gallons
Empty Weight	250 lbs.
Gross Weight	459 lbs.
Powerplant	Multi-rotary (3 x 40 HP)
Dimensions	9'L, 8.5'W, 4.2'H

XM SERIES AIRCRAFT IN DEVELOPMENT

All XM series VTOL aircraft use the same external configuration to ensure power-off stability and minimum drag. The XM-6 is a six passenger turbofan powered VTOL vehicle. It is suitable for executive or military applications where a larger payload, higher speed and longer range is needed. It will require additional development of the propulsion system but uses the same stabilizing system as the XM-5. The selling price of the XM-6 is comparable to that for a conventional turboprop powered executive aircraft.

XM-6 Specifications (Projected)

Passengers	6
Cruise Speed	325 MPH @ 65% power
Top Speed	390 MPH
Rate of Climb	4,300 fpm @ 75% power
Maximum Range	1,035 miles, 20 min. reserve
Payload	1,020 lbs.
Hover Ceiling	12,000 ft.
Operational Ceiling	36,000 ft.
Fuel Capacity	215 gallons
Empty Weight	1,490 lbs.
Gross Weight	3,800 lbs.
Powerplant	Augmented Turbofan
Dimensions	19'L, 18.5'W, 5.75'H

MOLLER CORPORATION
1222 RESEARCH PARK DRIVE
DAVIS, CA 95616
(916) 756-5069

A Machine to End War

by Nikola Tesla As Told To George Sylvester Viereck

Editor's note: Nikola Tesla, now in his seventy-eighth year, has been called the father of radio, television, power transmission, the induction motor, and the robot, and the discoverer of the cosmic ray. Recently he has announced a heretofore unknown source of energy present everywhere in unlimited amounts, and he is now working upon a device which he believes will make war impracticable.

Tesla and Edison have often been represented as rivals. They were rivals, to a certain extent, in the battle between the alternating and direct current in which Tesla championed the former. He won; the grat power plants at Niagara Falls and elsewhere are founded on the Tesla system. Otherwise the two men were merely opposites. Edison had a genius for practical inventions immediately applicable. Tesla, whose inventions were far ahead of the time, aroused antagonisms which delayed the fruition of his ideas for years.

However, great physicists like Kelvin and Crookes spoke of his inventions as marvelous. "Tesla," said professor A.E. Kennely of Harvard University when the Edison medal was presented to the inventor, "set the wheels going round all over the world...What he showed was a revelation to science and art unto all time.."

"Were we," remarks B.A. Behrend, distinguished author and engineer, "to seize and to eliminate the results of Mr. Tesla's work, the wheels of industry would cease to turn, our electric cars and trains would stop, our towns would be dark, our mills would be dead and idle."

Forecasting is perilous. No man can look very far into the future. Progress and invention evolve in directions other than those anticipated. Such has been my experience, although I may flatter myself that many of the developments which I forecast have been verified by events in the first third of the twentieth century.

It seems that I have always been ahead of my time. I had to wait nineteen years before Niagara was harnessed by my system, fifetnn years before the basic inventions for wireless which I gave to the world in 1893 were applied universally. I announced the cosmic ray and my theory of radioactivity in 1896. One of my most important discoveries - terrestrial resonance - which is the foundation of wireless power transmission and which I announced in 1899, is not understood even today. Nearly two years after I had flashed an electric current around the globe, Edison, Steinmetz, Marconi, and others declared it would not be possible to transmit even signals across the Atlantic. Having anticipated so many important developments, it is not without assurance that I attempt to predict what life is likely to be in the twenty-first century.

Life is and will ever remain an equation incapable of solution, but it contains certain known factors. We may definitely say that it is a movement even if we do not fully understand its nature. Movement implies a body which is being moved and a force which propels it against resistance. man, in the large, is a mass urged on by a force. Hence the general laws governing movement in the realm of mechanics are applicable to humanity.

There are three ways by which the energy which determines human preogress can be increased: *First*, we may increase the mass. This, in the case of humanity, would mean the improvement of living conditions, health, eugenics, etc. *Second*, we may reduce the frictional forces which impede progress, such as ignorance, insanity, and religious fanaticism. *Third*, we may multiply the energy of the human mass by enchaining the forces of the universe, like those of the sun, the ocean, the winds and tides.

The first method increases food and well-being. The second tends to bring peace. The third enhances our ability to work and to achieve. There can be no progress that is not constantly directed toward increasing well-being, peace, and achievement. Here the mechanistic conception of life is one with the teachings of Buddha and the Sermon on the Mount.

While I am not a believer in the orthodox sense, I commend religion, first, because every individual should have some ideal -religious, artistic, scientific, or humanitarian -to give significance to his life. Second, because all the great religions contain wise prescriptions relating to the conduct of life, which hold good now as they did when they were promulgated.

There is no conflict between the ideal of religion and the ideal of science, but science is opposed to theological dogmas because science is founded on fact. To me, the universe is simply a great machine which never came into being and will never end. The human being is no exception to the natural order. Man, like the universe, is a machine. Nothing enters our minds or determines our actions which is not directly or indirectly a response to stimuli beating upon our sense organs from without. Owing to the similarity of our construction and the sameness of our environment, we respond in like manner to similar stimuli, and from the concordance of our reactions, understanding is born. In the course of ages, mechanisms of infinite complexity are developed, but what we call "soul" or "spirit," is nothing more than the sum of the functionings of the body. When this functioning ceases, the "soul" or the "spirit" ceases likewise.

I expressed these ideas long before behaviorists, led by Pavlov in Russia and by Watson in the United States, proclaimed their new psychology. This apparently mechanistic conception is not antagonistic to an ethical conception of life. The acceptance by mankind at large of these tenets will not destroy religious ideals. Today Buddhism and Christianity are the greatest religions both in number of desciples and in importance. I belive that the essence of both will be the religion of the human race in the twenty-first century.

The year 2100 will see eugenics as firmly established. In past ages, the law governing the survival of the fittest roughly weeded out the less desirable strains. Then man's new sense of pity began to interfere with the ruthless workings of nature. As a result, we continue to keep alive and to breed the unfit. The only method compatible with our notions of civilization and the race is to prevent the breeding of the unfit by sterilization and the deliberate guidance of the mating instinct. Several European countries and a number of states of the American Union sterilize the criminal and the insane. This is not sufficient. The trend amoung eugenists is that we must make marriage more difficult. Certainly no one who is not a desirable parent should be permitted to produce progeny. A century from now it will no more occur to a normal person to mate with a person who is eugenically unfit than to marry a habitual criminal.

Hygiene, physical culture will be recognized branches of education and government. The Secretary of Hygiene or Physical Culture will be far more important in the cabinet of the President of the United States who holds office in the year 2035 than the Secretary of War. The pollution of our beaches as exists today around New York City will seem as unthinkable to our children and grandchildren as life without plumbing seems to us. Our water supply will be more carefully supervised, and only a lunatic will drink unsterilized water.

More die or grow sick from polluted water than from coffee, tea, tobacco, and other stimulants. I myself eschew all stimulants. I also practically abstain from meat. I am convinced that within a century coffee, tea, and tobacco will be no longer in vogue. Alcohol, however, will still be used. It is not a stimulant but a veritable elixir of life. The

This drawing accompanied the story and shows how the Tesla tower built on Long Island would have looked when completed. From it's apearance nobody would infer that it was to be used for the fantastic purposes set forth in the paper.

abolition of stimulants will not come about forcibly. It will simply be no longer be fashionable to poison the system with harmful ingredients. Benarr Macfadden has shown how it is possible to provide palatable food based upon natural products such as milk, honey, and wheat. I believe that the food which is served today in his penny restaurants will be the basis of epicurean meals in the smartest banquet halls of the twenty-first century.

There will be enough wheat and wheat products to feed the entire world, including the teeming millions of China and India, now chronically on the verge of starvation. The earth is bountiful, and where her bounty fails, nitrogen drawn from the air will refertilize her womb. I developed a process for this purpose in 1900. It was perfected fourteen years later under the stress of war by German chemists.

Long before the next century dawns, systematic reforestation and the scientific management of natural resources will have made an end of all devastating droughts, forest fires, and floods. The universal utilization of water power and its long-distance transmission will supply every household with cheap power and will dispense with the necessity of burning fuel. The struggle for existence being lessened, there should be development along ideal rather than material lines.

Today the most civilized countries spend a maximum of their income on war and a minimum on education. The twenty-first century will reverse this order. It will be more glorious to fight against ignorance than to die on the field of battle. The discovery of a new scientific truth will be more important than the squabbles of diplomats. Even the newspapers of our own day are beginning to treat scientific discoveries and the creation of fresh philosophical concepts as news. The newspapers of the twenty-first century will give a mere "stick" in the back pages to accounts of crime or political controversies, but will headline on the front pages the proclamation of a new scientific hypothesis.

Progress along such lines will be impossible while nations persist in the savage practice of killing each other off. I inherited from my father, an erudite man who labored hard for peace, an ineradicable hatred of war. Like other inventors, I believed at one time that war could be stopped by making it more destructive. But I found that I was mistaken. I underestimated man's combative instinct, which it will take more than a century to breed out. We cannot abolish war by outlawing it. We cannot end it by disarming the strong. War can be stopped, not by making the strong weak but by making every nation, weak or strong, able to defend itself.

Hitherto all devices that could be used for defense could also be utilized to serve for agression. This nullified the value of the improvement for purposes of peace. But I was fortunate enough to evolve a new idea and to perfect means which can be used chiefly for defense. If it is adopted, it will revolutionize the relations between nations. It will make any country, large or small, impregnable against armies, airplanes, and other means for attack. My invention requires a large plant, but once it is established it will be possible to destroy anything, men or machines, approaching within a radius of 200 miles. It will, so to speak, provide a wall of power offering an insuperable obstacle against any effective agression.

If no country can be attacked successfully, there can be no purpose in war. My discovery ends the menace of airplanes or submarines, but it ensures the supremacy of the battleship, because battleships may be provided with some of the required equipment. There might still be war at sea, but no warship could successfully attack the shore line, as tthe coast equipment will be superior to the armament of any battleship. I want to state explicitly that this invention of mine does not contemplate the use of any so-called "death rays." Rays are not applicable because they cannot be produced in requisite quantities and diminish rapidly in intensity with distance. All the energy of New York City (approximately two million horsepower) transformed into rays and projected twenty miles, could not kill a human being, because, according to a well-known law of physics, it would disperse to such an extent as to be ineffectual.

My apparatus projects particles which may be relatively large or of microscopic dimensions, enabling us to convey to a small area at great distance trillions of times more energy than is possible with rays of any kind. Many thousands of horsepower can thus be transmitted by a stream thinner than a hair, so that nothing can resist. This wonderful feature will make it possible, among other things, to achieve undreamed-of results in television, for there will be almost no limit to the intensity of illumination, the size of the picture, or distance of projection.

I do not say that there may not be several destructive wars before the world accepts my gift. I may not live to see its acceptance. But I am convinced that a century from now every nation will render itself immune from attack by my device or by a device based upon a similar principle.

At present we suffer from the derangement of our civilization because we have not yet completely adjusted ourselves to the machine age. The solution of our problems does not lie in destroying but in mastering the machine.

Innumerable activities still performed by human hands today will be performed by automatons. At this very moment scientists working in the laboratories of American Universities are attempting to create what has been described as a "thinking machine." I anticipated this development.

I actually constructed "robots." Today the robot is an accepted fact, but the principle has not been pushed far enough. In the twenty-first century the robot will take the place which slave labor occupied in ancient civilization. There is no reason at all why most of this should not come to pass in less than a century, freeing mankind to persue its higher aspirations.

And unless mankind's attention is too violently diverted by external wars and internal revolutions, there is no reason why the electric millenium should not begin in a few decades.

(Reprint of Nikola Tesla's last public work, written shortly before his death.)

UFO photographed by the U.S. military

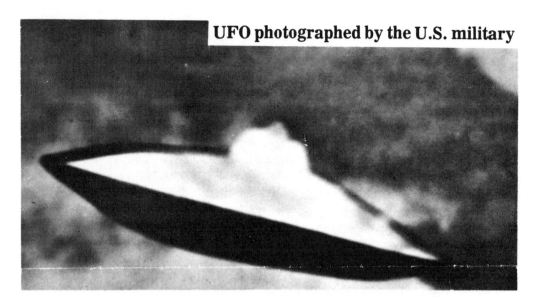

This photo was taken over the N.E. China Sea by a U.S. military photographer. The UFO, gold in color over its lower half, was described as having a high level of radiation emanating from its upper surface. After tracking the military aircraft for a short time, it reportedly vanished in mid-air.

Kenneth Arnold's sighting produced the term 'flying saucers' in 1947, for details of his and other people's UFO experiences at that time, see his book *The Coming of the Saucers*, co-authored with, and published by Ray Palmer, 1952.

Photograph taken by the pilot of an AVENA Airlines plane between Barcelona and Maiquetia, Venezuela. An interesting feature are the shadows cast by the plane and the object.

Reprinted from the September, 1927 issue of Science and Invention.

Gravity Nullified

Quartz Crystals Charged by High Frequency Currents Lose Their Weight

ALTHOUGH some remarkable achievements have been made with short-wave low power transmitters, radio experts and amateurs have recently decided that short-wave transmission had reached its ultimate and that no vital improvement would be made in this line. A short time ago, however, two young European experimenters working with ultra short-waves, have made a discovery that promises to be of primary importance to the scientific world.

The discovery was made about six weeks ago in a newly established central laboratory of the Nessartsaddin-Werke in Darredein, Poland, by Dr. Kowsky and Engineer Frost. While experimenting with the constants of very short waves, carried on by means of quartz resonators, a piece of quartz which was used, suddenly showed a clearly altered appearance. It was easily seen that in the center of the crystal, especially when a constant temperature not exceeding ten degrees C. (50 degrees Fahrenheit) was maintained, milky cloudiness appeared which gradually developed to complete opacity. The experiments of Dr. Meissner, of the Telefunken Co., along similar lines, according to which quartz crystals, subjected to high frequency currents clearly showed air currents which led to the construction of a little motor based on this principle. A week of eager experimenting finally led Dr. Kowsky and Engineer Frost to the explanation of the phenomenon, and further experiments showed the unexpected possibilities for technical uses of the discovery.

Some statements must precede the explanation. It is known at least in part, that quartz and some other crystals of similar atomic nature, have the property when exposed to potential excitation in a definite direction, of stretching and contracting; and if one uses rapidly changing potentials, the crystals will change the electric waves into mechanical oscillations. This *piezo electric* effect, shown in Rochelle salt crystals by which they may be made into sound-producing devices such as loud speakers, or reversely into microphones, also shows the results in this direction. This effect was clearly explained in August, 1925 *Radio News* and December, 1919 *Electrical Experimenter*. These oscillations are extremely small, but have nevertheless their technical use in a quartz crystal wave-meter and in maintaining a constant wavelength in radio transmitters. By a special arrangement of the excitation of the crystal in various directions, it may be made to stretch or increase in length and will not return to its original size. It seems as if a dispersal of electrons from a molecule resulted, which, as it is irreversible, changes the entire structure of the crystal, so that it cannot be restored to its former condition.

The stretching out, as we may term this strange property of the crystal, explains the impairment of its transparency. At the same time a change takes place in its specific gravity. Testing it on the balance showed that after connecting the crystal to the high tension current, the arm of the balance on which the crystal with the electrical connections rests, rose into the air. The illustration, Fig. 3, shows this experiment.

This pointed the way for further investigation and the determination of how far the reduction of the specific gravity could be carried out. By the use of greater power, finally to the extent of several kilowatts and longer exposure to the action, it was found eventually that from a little crystal, 5 by 2 by 1.5 millimeters, a non-transparent white body measuring about ten centimeters on the side resulted, or increased about 20 times in length on any side (see Fig. 4.) The transformed crystal was so light that it carried the whole apparatus with itself upwards, along with the weight of twenty-five kilograms (55 lbs.) suspended from it and floating free in the air. On exact measurement and calculation, which on account of the excellent apparatus in the Darredein laboratory could be readily carried out, it was found that the specific gravity was reduced to a greater amount than the change in volume would indicate. Its weight had become practically negative.

There can be no doubt that a beginning has been made toward overcoming gravitation. It is to be noted, however, that the law of conservation of energy is absolutely unchanged. The energy employed in treating the crystal, appears as the counter effect of gravitation. Thus the riddle of gravitation is not fully solved as yet, and the progress of experiments will be followed further. It is, however, the first time that experimentation with gravitation, which hitherto has been beyond the pale of all such research, has become possible, and it seems as if there were a way discovered at last to explain the inter-relations of gravity with electric and magnetic forces, which connection, long sought for, has never been demonstrated. This report appears in a reliable German journal, "Radio Umschau."

Fig. 1. The gravitation nullifier is shown in this illustration. The quartz crystal may be seen supporting a 55-pound weight. Dr. Kowsky is shown in a top coat because of the temperature at which the experiments were performed.

Fig. 3. This shows how the quartz crystal lost weight when subjected to the high frequency current. The original crystal was balanced on the scale.

Fig. 2. The schematic diagram of the experiment is shown in this illustration. The high frequency oscillator has been omitted for clearness.

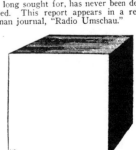

Fig. 4. This illustration shows the relative sizes of the crystal before and after the experiment. It is approximately twenty times its original length on any side.

Don't fail to see our next issue regarding this marvelous invention.

Nullified Gravity-A Hoax

Science and Invention for October, 1927

In our issue of September, on page 398, we ran an article entitled "Gravity Nullified,' with a subtitle "Quartz Crystals Charged with High Frequency Currents Lose Their Weight." At the end of tthe article we also ran a line, "*Don't Fail to See Our Next Issue Regarding This Marvelous Invention.*"

Those who were wise evidently must have had their suspicions aroused by the bottom line, and the wiser ones, if they inspected the main photograpg carefully, no doubt at once saw the hoax.

The article, which came to us from Germany, appeared originally in a German periodical as an April joke, but it was so excellent that we thought that we could take a little liberty with our own readers. The question remains as to how many of our readers were fooled.

If you look closely at the main illustration, which we reproduce herewith, you will observe that the article labeled "1" is nothing more nor less than a microphone with a resistance. "2" is a pair of head recievers, and "3" is an old time German telephone transmitter with a mouthpiece which, in this case, serves the practical jokester as a handle. Naturally the critical inspector of the picture must have wondered what two microphones and a pair of head receivers had to do with the Gravity Nullifier. Also the supporting wire does not even touch the ring on the weight. Anyhow, we ask our readers' indulgence for the little hoax, for which we hope to be pardoned because the article surrounding it seemed quite authoritative and contained really a lot of good science tending ot hide the hoax.

As a matter of fact, most of the statements are true, with the exception, of course, of those statements referring to the expanded crystal and to the loss of weight caused by the supposed high frequency currents.

There are so many wonderful things happening in science every day that he who would label anything as impossible may have to take his words back the next day. The real fact remains that gravity will be nullified sooner or later, and most likely by some such means as shown in the hoax in the September issue. That electricity and gravitation are closely allied no one doubts, and we would therefore not be surprised if even some of our more scientifically inclined readers, who did not pay close attention to some of the details, took the article as authentic.

Scientific hoaxes are no novelty. One of the most famous, which was not exposed as quickly as this one, appeared in no less than the New York *Sun*. At that time, in August, 1835, a certain professor was supposed to have submitted his report on a fantastic moon people to the *Edinburgh Journal of Science*, to which manuscript the New York *Sun* obtained the first rights, and the article ran consecutively over a period of time. These moon articles, written in a more or less scientific vein, aroused tremendous excitement, and the Moon Hoax was actually believed by thousands upon thousands of people at that time. Needless to say, the *Sun* afterwards exposed its hoax, but even though the newspaper did so, the hoax was still believed by thousands of individuals for years.

The moral is that we should not believe everything that we see, but do a little original thinking ourselves, because we may never know, otherwise, what are facts and what are not.

As a matter of interest to the editors, would like to hear from you as to your impression of the hoax article, and whether you believed it or not. This will give the editors a good basis for a compilation of interesting facts.

A Few Notes on the Articles

In this short commentary we will examine the apparent inconsistencies in the preceding articles and the scientific evidence supporting the first article reported in a recently published experiment by J.G. Gallimore which follows this commentary.

It makes one wonder why such a magazine as Science and Invention would put their credibility on the line by publishing such a hoax. It is possible that they intended the sensational article to boost their sales much as the "fantastic moon people" article did for the New York Sun (this series of articles boosted the paper's circulation to the highest of any paper in the world at the time). However, one must wonder, in light of current knowlege of harmonic math and the incredible properties of crystals, whether or not the real hoax is not Science and Inventions retraction. Also, J.G. Gallimore claims to have successfully reproduced this experiment as reported in the Planetary Association for Clean Energy newsletter, volume 2, Numbers 4 & 5, Febuary 1981 (this article follows).

Can this be called a hoax because the equipment bears a resemblence to certain mundane articles? It is commonplace for engineers to build projects from any scrap materials that they could use to prove the viability of their project. In light of this, could those items be microphones, a German telephone transmitter, a head receiver, etc.? Quite possibly those items may have been the raw materials that they used for the experiment, and the "scrap" material that they used no longer functions the way it did originally. Is it true that high frequency currents may produce anti-gravity effects? Referring to the article "Nullified Gravity--A Hoax": "As a matter of fact, most of the statements are true, with the exception, of course, of those statements referring to the expanded crystal and to the loss of weight caused by the supposed high frequency currents." If the October, 1927 article is indeed a hoax, then this may be similar to the Orwellian "Newspeak", and the very thing that they deny so vehemently is actually the truth. Actually, for the publication to take this stand would be an insult to the work of Nikola Tesla who experimented with the anti-gravitic effects of high frequency currents of high potentials and found them to "contain great promise" (see Bib.).

A number of inconsistencies have been revealed in the article "Nullified Gravity--A Hoax" through the evolution of our technology since its appearence in September, 1927. One inconsistency involves the digital reprocessing of the photograph that is in the article "Gravity Nullified". Digitizing the photo reveals that the ring actually does seem to touch the supporting wire.

Another interesting point is the admonition:

"There are so many wonderful things happening in science every day that he who would label anything as impossible may have to take his words back the next day. The real fact remains that gravity will be nullified sooner or later, and most likely by some such means as shown in the hoax in the September issue. That electricity and gravitation are closely allied no one doubts..."

And why choose a frequency within the range that they selected to conduct the experiments? That "magic" frequency just happens to be within the same range used by other anti-gravity and "free energy" researchers. Which leads one to suspect that they knew exactly what they were doing when they performed the experiment. And why would a hoax have so much apparently valid experimental data? Usually scientific hoaxes have intrinsic inconsistencies concerning the experimental parameters that are used, and it is those very inconsistencies that reveal the hoax. This contrasts to the body of the article "Gravity Nullified".

Also, in the article "Gravity Nullified," in the second paragraph, first line, "The discovery was made about six weeks ago in a newly established central laboratory..." This statement was made in the September issue, which would have put the article's appearence sometime in July. This would have been a bit late for an April fool's joke, as is stated: "The article, which came to us from Germany, appeared originally in a German periodical as an April joke...", in the article above. It seems curious that a greater amount of inconsistency appears in the text debunking the original article than in the original article itself.

In conclusion, the only way the world will ever know what is really going on is to repeat the experiment as outlined in the original article, to see if the observed Anti-Gravity effects actually exist.

W.P. Donavan & D.H. Childress

Anti-Gravity Properties of Crystalline Lattices

In the summer of 1927, two scientists, **Kowsky** and **Frost**, in Poland noted specific anti-gravity properties of crystals. They were pursuing some discoveries in piezo-electricity made by **Meissner** of Telefunken, whereby it was found that crystals could lose their transparency and change their specific gravity at the same time.

By the oscillations of radio transmitters of several kilowatts, at protracted exposure, Kowsky and Frost managed to include an eight hundred percent volume increase to a clear crystal. The small, lightened crystal carried the apparatus which oscillated it as well as a weight of twenty five kilograms suspended from it, floating free at a height of about two meters above the floor of a laboratory.

Shortly after this discovery, reports and photographs of the tests were published in the German journal, Radio Umschau and in Science and Invention (September, 1927 issue).

Those published reports permit a definition of the phenomena in today's terminology.

An optical grade quartz crystal 5x2x1.5 mm of defined crystal lattice was piezoelectrically overloaded with a resulting opaqueness, a growth in volume and a structural change along with specific gravity change. The crystal was reported to increase dimensions along one side of two thousand percent (volume increase of 800%). Its weight of approximately one ounce was reduced by an unknown amount during the increase in volume. When electrically excited to lift itself, the crystal was capable also of lifting an additional eight hundred and eighty ounces. This lift occurred when the crystal was subjected to vertical oscillation via direct electrode contacts, and transverse oscillation via non-attached electrodes broadcasting radiation with the crystal interposed between them.

Radio Frequency Emissions and Magnetostriction of Mass

Magnetostrictive masses emit heat and undergo dimensional changes on a temporary basis when exposed to a varying magnetic field. The molecular alignment of the mass with the field of current induces mechanical pressures that cause a distortion or dimensional change. Normally such physical changes have been assumed to be temporary or of unimportant plasticity. Certain non-magnetic substances like dielectric crystals also react to an imposed magnetic field with molecualr re-alignment.

The re-alignment causes a crystal distortion in one direction, and with alternating current fields, oscillation occurs. Such is the piezo-electric phenomenon. The angle of turn of the molecule on its axis is proportional to the "strength" of the induced magnetic field until a limit of saturation is reached: "weber angle", or maximum distortion potential of the dielectric.

If additional power is appled to create a still stronger magnetic field, molecules that happen to exceed weber angles are wrenched away to migrate along the field path, to form bonds at new positions of equilibrium. The displaced and re-positioned molecules are termed "deflexions", or displaced ions (**Maxwell, Jeans**, 1916).

The magnetic susceptibility of a substance varies inversely as the temperature (Curies Law). This experiment potentially justifies a "K", or a susceptibility enhancement by the 'freeze storage' of all new re-positioning ions, and consequent stability in new positions. Ion bonds form slowly in a dielectric heated by intense magnetic field changes, known as 'inductance heating'. Cooling of the dielectric by air currents arond the dielectric which draws off heat allows the dielectric to escape destruction by melting, brittle fracture, or other heat-caused affects.

The migration of displaced ions is to a surface area of the dielectric, where the heat sink phenomena allows a re-bonding temperature.

Known research in electric action versus dielectrics leads to other supporting information about the physical phenomenon.

- **Helmholtz**: The value of "K" changes in a dielectric when it is subjected to distortion. (K equals the dielectric constant of that mass..)
- **Maxwell**: With displacement, the density of the medium (crystal structure) is changed so that its molecular structure is changed; as is its "K".
- The K of quartz depends on the direction of the imposed magnetic field "relative" to the crystal axis. A vertical K of quartz is 4.55, and horizontal K is 4.49 where K is a reaction to the earth's field.
- Magnetic conduction in a dielectric is altered as if the properties of the medium were altered during conduction by a change of the dielectric constant of the mass itself.

So far the phenomenon appears not to be rejected by known physical actions. About the phenomenon itself, a brief theoretical model may be postulated.

This may be a stress model of mass where changes of internal stress induce 'deformation of mass'. Thus the model suggests a 'two-part' investigation; (1) the stress model, (2) the later physical phenomena produced as a product of distortion, and the physical performance relative to change.

The Stress Model

The electric force between charged particles is independent of the masses of energies of the particles, and depends only on their charge; whereas, the gravitational force is proportional to the masses themselves. Since in special relativity mass and energy are related by $E=mc^2$, the 'strength' of the gravitational field increases as the energies of the virtual particles increase.

An artificially induced increased 'stress' in mass increases the energy of both virtual particles and gravity.

In an electron flow such as common electricity along a conductor, the 'pressure' of the flow affects the mass of the conductor by several methods:

1. free electrons are displaced by induced energy.

2. torque from electrical action is appied to the mass.

3. stresses are induced within the mass.

4. compression is induced within the mass.

5. structure bonds are affected by such imposed stresses.

6. ion orbital structure is relative to the induced energy; greater energy produces energy absorption with smaller orbits, higher velocities, (packing fraction).

Electric/Dielectric Combinations

The electrostatic attracttion of one object to another depends on charge, shape, and surface area; but the magnetic attraction to a fragment of dielectric is a molecular phenomenon 'independant of shape', but not surface area. A non-magnetic body will be components of/or magnetic particles when a magnetic field is imposed; ie. an attraction. Finally, a magnetic field will exist in a dielectric after/when an induced field changes, or is no longer imposed. So it may be assumed that an intense magnetic field is the one force which is capable of externally affecting the dielectric molecular axis change.

Further Investigations

Alternating currents produce heat, and a magnetic field, in a dielectric to a depth proportional to the square root of the oscillation period; and to the applied strength. A magnetic particle, or ion, is capable of re-positioning, where 'all' mass particles are also susceptible. Unlike metals, a dielectric 'acts' as if it conducts one hundred percent of any imposed field. This is the single most important difference.

The molecular phenomenon may occur 'only' in a dielectric mass, and not in a metal. The magnetic conduction proportional to field depth with a strength sufficient to dislodge ions eliminates metals (exception: Bismuth) due to skin affect. The dielectric conduction of one hudred percent of imposed fields "throughout" the mass allows the 'total' mass to be involved, eliminating skin affect. So, it will affect all crystalline lattice structures (therefore metals affected), however the optimum solution for maximum affect of this phenomenon may reside within the electrical characteristics of dielectrics.

Magnetic Induction Currents

Magnetic induction postulated as a solenoidal induction throughout the field in the interior of the mass (all points equal) can occur in a dielectric, but not in a metal. Magnetostriction of dielectrics: there is an expansion of mass proportional to the induced strains (internal) to release pressure. This is a known, accepted phenomenon. There are 'diamagnetic' currents induced in the same crystal (opposite to magnetic) about which little is known, but which has been photographed at Gallimore Labs. Such currents are always found in 'stressed masses'.

Crystal Expansion Confirmed

Crystal expansion was examined from actual replication of the Kowsky and Frost experiment.

The crystal will have intense internal strains, generally 'only' in the direction of applied fields producing expansion, and diamagnetic currents of unknown effects or phenomena. (Many phenomena were detected).

The revised theory of phenomenon is stated as: The molecular motion and reaction of mass to intense magnetic fields may change the structure of the (dielectric) mass if the imposed field strength exceeds the force needed to rotate fixed molecules past Webers angle, where the result would be a dislocation of the molecule from the mass structure. Given this field strength, it is almost certain that the re-positioning of molecules will change the normal lattice structure, and will be accompanied by permanent expansion of the dielectric along the vector of the imposed field.

Such diamagnetic currents as exist will be intense, and could produce a host of phenomena. The proposed 'Anti-gravity" phenomena. The proposed 'Anti-gravity' phenomena fall within an 'acceptable' but not proven phenomena at the present time.

It is noted that from the **Chicago College on Gravity Research** that a 60 Hz. alternating current imposed on a solenoid when placed under an aluminum plate, will cause the plate to heat, as well as 'lift' upwards as much as eleven inches. Such a lifting effect cannot come from magnetic actions, but may come from molecular actions, and the little known diamagnetic currents.

Since magnetic fields in alternating currents become stronger as the frequency increases, higher frequencies are found more efficient in producing the 'stress fields' producing lift phenomena.

Of interest is that one dielectric has been shown to 'fall' more slowly under 'natural' conditions than any mass shoud fall. It is unknown whether aluminum silicate reacts to existent low intensity magnetic fields, or whether it has an excessive diamagnetic current capability occuring naturally.

Water Absorption/Emission

The expanded crystal has been found to be both effervescent and deliquescent. This is unusual in a single mass; to absorb and release water like a sponge where the material (silicon dioxide) is neither an absorber or emitter prior to change of the mass structure, a degeneration of the structure is seen after one water cycle, and is apparently not repeatable.

The Kowsky and Frost experiment was reported to have a visual sighting of air currents flowing around the crystal when under electrical excitation. It is a fact that the air currents so described are a reality, but are not known to exist by crystallographers, unless they have considerable experience in electrical testing. Likewise, electrical testers and engineers are not likely to have witnessed this. It is here noted that air currents have been found around excited (oscillating) crystals, but only when a frequency band of one hundred kilocycles to four hundred eighty kilocycles is utilized. This is a further verification of actual research being in the frequency range specified, and of a true research sighting being transmitted.

Electricity Produced

An expanded lattice crystal has been found to produce a remarkable phenomenon: when an 'imbalance' occurs by stress changes in a 'stress balanced' crystal, electricity is produced.

A crystal 'grown' in an unbalanced state will 'convert one hundred percent of all radiation reaching it to electricity'.

Lift Factor

the following equation is only generalized, and its veracity should be questioned. It may be a guideline of potential results. The resulting values are indicative within limits of what can be expected experimentally.

$$\frac{\text{Force applied in watts} \times \text{mass in Kg} \times \text{Expansion \%}}{\text{Frequency}/7770 \qquad\qquad\qquad 100} = \text{Kg lift}$$

Example: $2000 \text{ watts} \times \dfrac{(5\text{kg} \times 300\%)}{100} = \dfrac{2000 (150)}{777 \text{ kc}/7770 \quad 100} = \dfrac{3000}{100} = 30 \text{ kg lift}$

Mass Structure and Potential Collapse

All mass is susceptible to change. Dielectrics by having an organized molecular structure are subject to massive change through force applied. It is considered a phenomenon where lattice structure is expanded, and re-formed to a new related structure by energy; and that the structure is now a "storage medium" of great energies by strain locked in structure.

The stability of the medium, or rather the changed medium, is now questioned, as well as its life span. A sudden sharp blow or even chemical activity may "detonate" or collapse the new structure with great release of energy. This potential is seen at the present time to be both real and hazardous. Should the crystal mass be capable of sudden collapse, it could take one of two forms; sudden disintegration to a powder state, or detonation with a massive release of energy, perhaps similar to atomic conversion of mass to energy.

Self Contained Lifting Device

The subject mass utilized in this research has been quartz dielectrics. Quartz, unlike many substances, does not shear easily; but has a conchoidal fracture. Providing it did shear, then he lattice structure could be pried open at selected locations, and slabs of the expanded variety utilized in different applications. Because of the energy storage phenomena, it is assumed it cannot be 'sawed' as the shock potential is high, yet proportional to the degree of the crystalline lattice expansion.

The crystal itself will resemble plastic foam in weight and rigidity. Perhaps it could be sliced by a laser or electron beam.

Very little power is required to oscillate the substance for high lift. This and the weight needed to supply that power allows a fully contained device to be a reality. Power applied as frequency would have six basic contacts regardless of design, size, or aerodynamic shape; ie. left side, right side, front, rear, top and bottom.

With solenoid controls, the full range of flight could be obtained; forward, right, up, down, reverse. The control would be by reversing polarities of a given area of surface section to provide the desired result. Each lift/control section would be electrically isolated in a smooth surface design by interposing non-expanded dielectric strips beween sections. Such skin or the dielectric isolators could provide shape, rigidity, and supporing design.

Such dielectrics as ceramics display temperature resistance, and could be included as "skin." However, since speed is fully controllable, there should be no need for heat buildup; simply reduce speed.

Reverse Phenomenon

In trying to validate mass structure cahnge as proposed, the 'reverse' method of gravitational emission (ie. "absorption") was used.

If a mass may "produce" radiation under coercion, then it may also be susceptible to that same radiation when exposed thus providing a "reverse phenomenon."

The lattice structure of a dielectric has been proposed as a storage medium of immense energies, when the energy applied produced a 'Deflexion' change (deflected ions) or structural stresses of great magnitude. A first discovery was that by utilizing a 'new' means of electrical excitement, a "commercial" process of Deflexion crystals was realized. In further research, the 'reverse' phenomenon indicates that 'all' dielectrics having 'any' stress components may be susceptible to gravity radiation.

Discovered in 1974, the reverse phenomenon allowed for an on-time gravity monitor where a dielectric with a known stress component was seen to change proportional to the acting gravitational intensities. Later, in 1978 a new detector was discovered, the difference being that a 'general' state of stress was utilized here as opposed to a known 'finite' stress.

(J. G. Gallimore)

A Few Notes on the Article

This article appeared in the February, 1981 issue of the Planetary Association for Clean Energy newsletter. It contains quite a bit of valuable information. It also has questionable information, which we shall review.

The reference articles that are referred to in Radio Umschau and Science and Invention was followed up in the October, 1927 issue of Science and Invention. This article is titled "Nullified Gravity-A Hoax" and completely disproves both the article in Radio Umschau and the one in the September, 1927 issue of Science and Invention.

The question arises: Why disprove all this early data if it is valid? And if it is not valid, why has J. G. Gallimore compiled all that data just to make an elaborate hoax look convincing? Obviously the answer must wait until further correlation from other researchers on this field has been published.

For now it would seem to be prudent to examine the work at hand. One questionable point is the equation. If it is examined closely, some simple errors in multiplication will be noted. Two corrected versions of this equation will be shown.

Equation 1 is the original as seen in the article. Equation 2 is a corrected version based on the assumption that Equation 1 is wrong and the product would also be incorrect. Equation 3 assumes that the product is correct (it may have been the actual results of laboratory testing) and the equation is wrong. Equation 4 is a simpified version of the original based upon equation 3. Here are the equations:

Equation 1:

$$\frac{\text{Force applied in watts} \times \text{mass in Kg} \times \text{Expansion \%}}{\text{Frequency}/7770 \qquad\qquad 100} = \text{Kg lift}$$

Example: $2000 \text{ watts} \times \frac{(5kg \times 300\%)}{777 kc/7770} = \frac{2000 \times (150)}{100 \quad 100} = \frac{3000}{100} = 30 \text{ kg lift}$

Equation 2:

$$\frac{\text{Force applied in watts} \times \text{mass in Kg} \times \text{Expansion \%}}{\text{Frequency}/7770 \qquad\qquad 100} = \text{Kg lift}$$

Example: $2000 \text{ watts} \times \frac{(5kg \times 300\%)}{777 kc/7770} = \frac{2000 \times (15)}{100 \quad 100} = \frac{300}{100} = 3 \text{ kg lift}$

Equation 3:

$$\frac{\text{Force applied in watts} \times \text{mass in Kg} \times \text{Expansion \%}}{\text{Frequency}/7770 \qquad\qquad 10} = \text{Kg lift}$$

Example: $2000 \text{ watts} \times \frac{(5kg \times 300\%)}{777 kc/7770} = \frac{2000 \times (15)}{100 \quad 10} = \frac{300}{10} = 30 \text{ kg lift}$

Equation 4:

$$\frac{\text{Force applied in watts} \times \text{mass in Kg} \times \text{Expansion \%}}{\text{Frequency}/777} = \text{Kg lift}$$

Example: $2000 \text{ watts} \times \frac{(5kg \times 300\%)}{777 kc/777} = \frac{2000 \times (15)}{1000} = 30 \text{ kg lift}$

Another point of dissention concerns this paragraph:

"Since magnetic fields in alternating currents become stronger as the frequency increases, higher frequencies are found more efficient in producing the 'stress fields' producing lift phenomena."

Upon examination of the equation, the reverse would be true. A higher frquency would produce a higher divisor, and thus a lower efficiency. If this paragraph is right, then the equation is dead wrong (or vice-versa). If the equation is right, then efficiency would surpass 100% at low frequencies and result in more mechanical power output than can be accounted for in consideration of power input in watts.

In closing, it would seem that further corroboration of the research data is needed to clear up these points, possibly involving a second experiment to duplicate any observed anti-gravity effects.

W.P. Donavan

ELEVEN THINGS THAT NASA DISCOVERED ABOUT THE MOON THAT YOU NEVER KNEW.

"It seems much easier to explain the nonexistence of the moon than its existence." -NASA scientist Dr. Robin Brett

1. The Puzzle of the Moon's Origin: Scientists have generally offered three major theories to account for the moon in orbit around our planet. All three are in serious trouble, but the least likely theory emerged from the Apollo missions as the favorite theory. One theory was that the moon might have been born alongside the earth out of the same cosmic cloud of gas dust about 4.6 billion years ago. Another theory was that the moon was the earth's child, ripped out the Pacific basin, possibly. Evidence gathered by the Apollo program indicates though that the moon and the earth differ greatly in composition. Scientists now tend to lean toward the third theory- that the moon was "captured" by the earth's gravitational field and locked into orbit ages ago. Opponents of the theory point to the immensely difficult celestial mechanics involved in such a capture. All of the theories are in doubt, and none satisfactory. NASA scientist Dr. Robin Brett sums it up best: "It seems much easier to explain the nonexistence of the moon than its existence."

2. The Puzzle of the Moon's Age: Incredibly, over 99 percent of the moon rocks brought back turned out upon analysis to be older than 90 percent of the oldest rocks that can be found on earth. The first rock that Neil Armstrong picked up after landing on the Sea of Tranquility turned out to be more than 3.6 billion years old. Other rocks turned out to be even older; 4.3, 4.5, 4.6, and one even alleged to be 5.3 billion years old! The oldest rocks found on earth are about 3.7 billion years old, and the area that the moon rocks came from was thought by scientists to be one of the youngest areas of the moon! Based on such evidence, some scientists have concluded that the moon was formed among the stars long before our sun was born.

3. The Puzzle Of How Moon Soil Could Be Older Than Lunar Rocks: The mystery of the age of the Moon is even more perplexing when rocks taken from the Sea of Tranquillity were young compared to the soil on which they rested. Upon analysis, the soil proved to be at least a billion years older. This would appear impossible, since the soil was presumably the powdered remains of the rocks lying alongside it. Chemical analysis of the soil revealed that the lunar soil did not come from the rocks, but from somewhere else.

4. The Puzzle of Why the Moon "Rings" like a Hollow Sphere When a Large Object Hits it: During the Apollo Moon missions, ascent stages of lunar modules as well as the spent third stages of rockets crashed on the hard surface of the moon. Each time, these caused the moon, according to NASA, to "ring like a gong or a bell." On one of the Apollo 12 flights, reverberations lasted from nearly an hour to as much as four hours. NASA is reluctant to suggest that the moon may actually be hollow, but can otherwise not explain this strange fact.

THE ANTI-GRAVITY HANDBOOK   99

5. The Puzzle of the Mystifying Maria of the Moon: The dark areas of the moon are known as maria (seas, as this is what they looked like to early astronomers- dried-up seas). Some of these maria form the familiar "man-in-the-moon" and are, strangely, located almost entirely on one side of the moon. Astronauts found it extremely difficult to drill into surface of these dark plainlike areas. Soil samples were loaded with rare metals and elements like titanium, zirconium, yttrium, and beryllium. This dumbfounded scientists because these elements require tremendous heat, approximately 4,500 degrees fahrenheit, to melt and fuse with surrounding rock, as it had.

6. The Puzzle of the Rustproof Iron Found on the Moon: Samples brought back to earth by both Soviet and American space probes contain pure iron particles. The Soviets announced that pure iron particles brought back by the remote controlled lunar probe Zond 20 have not oxidized even after several years on earth. Pure iron particles that do not rust are unheard of in the scientific world (although there is a solid iron pillar of unknown age in New Delhi, India, that has also never rusted, and no one knows why).

7. The Puzzle of the Moon's High Radioactivity: Apparently, the upper 8 miles of the moon's crust are surprisingly radioactive. When Apollo 15 astronauts used thermal equipment, they got unusually high readings, which indicated that the heat flow near the Apennine Mountains was rather hot. In fact, one lunar expert confessed: "When we saw that we said, 'My God, this place is about to melt! The core must be very hot.' " But that is the puzzle. The core is not hot at all, but cold (in fact, as was assumed, it is a hollow sphere). The amount of radioactive materials on the surface is not only "embarrassingly high" but, difficult to account for. Where did all this hot radioactive material (uranium, thorium, and potassium) come from? And if it came from the interior of the moon (unlikely), how did it get to the moon's surface?

8. The Puzzle of the Immense Clouds of Water Vapor on the Dry Moon: The few lunar excursions indicated that the moon was a very dry world. One lunar expert said that it was "a million times as dry as the Gobi Desert." The early Apollo missions did not find even the slightest trace of water. But after Apollo 15, NASA experts were stunned when a cloud of water vapor more than 100 square miles in size was detected on the moon's surface. Red-faced scientists suggested that two tiny tanks, abandoned on the moon by U.S. astronauts, had somehow ruptured. But the tanks could not have produced a cloud of such magnitude. Nor would the astronauts' urine, which had been dumped into the lunar skies, be an answer. The water vapor appears to have come from the moon's interior, according to NASA. Mists, clouds and surface changes have allegedly been seen on the moon over the years by astronomers. For instance, six astronomers in the last century have claimed to have seen a mist which obscured details in the floor of the crater Plato. Clouds on the moon are extremely odd, because the moon's supposed small gravity (one sixth of the earth's, claim many conventional scientists and NASA) could not hold an atmosphere or have any clouds on it at all.

9. The Puzzle of the Glassy Surface on the Moon: Lunar explorations have revealed that much of the moon's surface is covered with a glassy glaze, which indicates that the moon's surface has been scorched by an unknown source of intense heat. As one scientist put it, the moon is "paved with glass." The experts' analysis shows this did not result from massive meteor impactings. One explanation forwarded was that an intense solar flare, of awesome proportions, scorched the moon some 30,000 years or so ago. Scientists have remarked that the glassy glaze is not unlike that created by atomic weapons (the high radiation of the moon should also be considered in light of this theory).

10. The Puzzle of the Moon's Strange Magnetism: Early lunar tests and studies indicated that the moon had little or no magnetic field. Then lunar rocks proved upon analysis to be strongly magnetized. This was shocking to scientists who had always assumed that the rocks had "some very strange magnetic properties...which were not expected." NASA can not explain where this magnetic field came from.

11. The Puzzle of the Mysterious "Mascons" Inside the Moon: In 1968, tracking data of the lunar orbiters first indicated that massive concentrations (mascons) existed under the surface of the circular maria. NASA even reported that the gravitational pull caused by them was so pronounced that the spacecraft passing overhead dipped slightly and accelerated when flitting by the circular lunar plains, thus revealing the existence of these hidden structures, whatever they were. Scientists have calculated that they are enormous concentrations of dense, heavy matter centered like a bull's-eye under the circular maria. As one scientist put it, "No one seems to know quite what to with them."

NASA, THE MOON AND
ANTI-GRAVITY

by David H. Childress

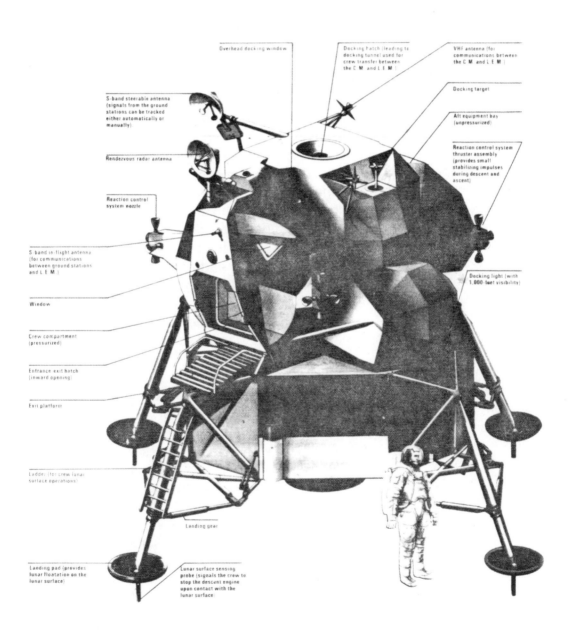

According to conventional science, the moon has only one-sixth of the Earth's gravity. Sir Isaac Newton formulated the Law of Universal Gravitation in 1666 which led to this conclusion. The famous law states that the gravitational pull of one body on another body depends on the product of the masses of the two bodies. Therefore, a planet such as the earth will attract another object with this force. The further out in space one goes, the less attraction is exerted on it by the body.

That the moon's gravity is one sixth of the earth's has been assumed for centuries, though there is now evidence that this not the case. William L. Brian II, a Nuclear Engineer from Oregon State University investigated what he calls a "NASA cover-up" in his 1982 book entitled: **MOONGATE: SUPPRESSED FINDINGS OF THE U.S. SPACE PROGRAM (THE NASA-MILITARY COVER-UP)."**

Brian centers his argument for a cover-up on the so-called "neutral point" between the earth and the moon. This neutral point, and all gravitational bodies have them, is the point where a space vehicle enters the predominant attractive zone of the moon's gravity. It is the region in space where the earth's force of attraction equals the moon's force of attraction. Since the moon is smaller and supposedly has a lesser surface gravity, the neutral point should be quite close to the moon. If the moon has one sixth of the earth's gravity, the neutral point is calculated to be about nine-tenths of the distance between the earth and the moon. The average distance to the moon is about 239,000 miles; hence this places the neutral point aproximately 23,900 miles from the moon's center.

Throughout the fifties and sixties, the neutral point between the earth and the moon was given over and over again as between 22,078 and 25,193 miles from the center of the moon. These figures, merely logical guesses from trained scientists, were based upon Newton's Law of Universal Gravitation. However, only by observing falling or orbiting bodies in the moon's vicinity could the actual neutral point distance, hence the moon's true gravity, be determined.

Using the above neutral point and assumed gravity for their calculations, the U.S. and the Soviet Union began to send space probes to the moon in the late fifties. They met with miserable failure. The Russians were the first to launch a successful lunar probe, Luna 1, on January 2, 1959. It flew within 4,660 miles of the moon and broadcast back information before continuing on into deep space. The U.S. made three unsuccessful attempts before achieving a fly-by of 37,000 miles from the surface some months after Luna 1 with their Pioneer 4. The Russian's Luna 2 became the first space probe to hit the moon and Luna 3 circled the far side of the moon, approached within 4,372 miles, and sent back photos of the far side. Strangely, Russian moon exploration came to a four year stop after these successes. Furthermore, the Russian's were intensely secretive about the data they collected.

The American efforts were almost laughable at first. The Ranger Space Probes were designed to hard land on the moon with seismometers in spherical containers designed to withstand the impacts on the moon. Ranger 3, launched on January 26, 1962, missed its target completely and went into a solar orbit. Ranger 4 hit the moon but did not send back any useful information. Ranger 5 missed the moon by 450 miles and then the effort was put off for two years while the entire program was reorganized-- something was wrong with their calculations!

Ranger 6, launched on January 30, 1964, allegedly had its electrical system burn out in flight and no pictures were sent. Subsequent Ranger probes were more successful. The Russians reactivated their space probes, but their Luna 5; launched on May 9, 1964; crashed at full speed on the moon, when it was intended to make a soft landing. Luna 6 utterly missed the moon, and Luna 7 crashed on the moon's surface when its retro-rockets supposedly fired too soon. Luna 8 also crashed on the moon, but Luna 9 became the first probe to successfully soft land on the moon.

Missions became more successful after this, and Brian alleges that this is because the Soviets and Americans had been able to recalculate the neutral point and correct gravity of the moon from the many failures (and the few successes).

The strangest thing to come out of the reanalyzing of "observing bodies fall and orbit the moon" was for NASA to come up with a new neutral point between the earth and the moon. And this is the key to "a NASA cover-up", according to Brian.

The July 25, 1969 issue of Time magazine stated that the neutral point was 43,495 miles form the center of the moon. Werner von Braun in the 1969 edition of "History of Rocketry & Space Travel", said that the neutral point was 43,495 miles from the center of the moon. Other reliable sources, obtaining their information from NASA, claimed that the neutral point was between 38,000 and 43,495 miles from the center of the moon. The pre-Apollo distances were given as 20,000 to 25,000 miles from the center of the moon. NASA, it appears, had recalculated the neutral point, which would indicate that the moon's gravity is not one sixth, as Newton had stated. We invite readers to research these figures for themselves.

According to Brian, if the neutral point of the moon's gravity is 43,495 miles from the moon, then the gravitation of the moon is 64% of the Earth's surface gravity, not one-sixth or 16.7% as predicted by Newton's Law of Universal Gravitation!

Bizarrely, NASA and the status-quo of science and government continue to allude that the moon's gravity is one-sixth that of the earth's, representative of a neutral point less than 25,193 miles from the moon. Brian goes on to say that if the neutral point were more like 52,000 miles instead of 43,495 miles from the moon, the moon's surface gravity would be identical to the earths!

Brian's data dwells heavily on inconsistent and contradictory information released through NASA, indicating an official cover-up. Brian's book goes on to list other evidence that the moon has a higher gravity than previously assumed; in fact, a gravity nearly equal the earth's! He points out the flight times of the Apollo spacecraft are inconsistent and much faster than if the moon had only one-sixth gravity, as the vehicle would continue to decelerate until it reached the neutral point, at which point it would begin to accelerate again as it became pulled by the moon's gravity.

This shot, taken on the Moon, shows spacecraft landing near dome-shaped building.

Copyright © 1959 by Howard Menger

AVRO Disc, 1955 Portrayal (Official U.S. Air Force Photo)
Each year, from 1954-1959, Air Force released this 'artists conception' of their 'secret', jet propelled flying saucer being built in Canada. Finally, in mid-1959 the finished product was tested.

Assuming that Brian is right about a higher surface gravity on the moon, then the ramifications of the high gravity on the fuel requirements of the lunar descent and ascent vehicle of the manned Apollo program to the moon are horrific! The Apollo launch rocket which sent men to the moon stood 363 feet tall and weighed 6,400,000 pounds. The payloads of the Lunar Module are correct, assuming that the moon had only one-sixth gravity. However, under a high lunar gravity the Lunar module would have had to have been nearly as large as a Titan 2 rocket which weighs 330,000 pounds and was 103 feet tall (the actual Lunar module, according to NASA, weighed 33,200 pounds)! The startling conclusion is that if men really landed on the moon in high lunar gravity conditions, it was not done with rockets!

These amazing conclusions, backed up by many scientists and using NASA data itself, raise a number of questions: Why did the Russians apparently pull out of the space race when they were hot on the trail of putting a man on the moon? How did the United States succeed when rockets would clearly not work in the high lunar gravity conditions? What was the military's involvement in top secret research which led to the successful moon landings? Did we even go to the moon as claimed? One publication, also by a former NASA employee, uses the same information from NASA used by Brian to "prove" that we never even made the trip! (This book is **"We Never Went To The Moon"**, by Bill Kaysing, 1981, Desert Publications, Cornville, Arizona).

Brian goes on to offer more "proof" that the moon's gravity is nearly equal to that of earth's, and that NASA staged the whole show in order to fake a one-sixth gravity, presumably to hide the fact that they had to use the ultra-secret "anti-gravity" devices that were what really powered the Lunar Module.

Consider these "facts":

In one sixth gravity, a 180 pound man would weigh a mere 30 pounds. Writers had programmed the public to expect athletic feats of a spectacular nature when astronauts explored the moon. In the November 1967 issue of Science Digest an article entitled, **"How To Walk on the Moon"** was printed that predicted that men would be able to make 14-foot slow-motion leaps, perform backflips and other gymnastics like professionals, and be able to easily move up ladders and poles with their arms. An astronaut, even in a cumbersome suit could jump six times higher on the moon than earth.

Even though the alleged weight of the spacesuits and backpacks of the astronauts was 185 pounds, the total combined weight of a 185 pound astronaut and his suit would be only 62 pounds. This is still only one third of the astronaut's weight. Therefore, the astronauts should have been able to jump vertically far higher than they could on Earth without any burden. Brian believes that an average man can jump about 18 inches high without a run. Basketball players routinely jump over three feet high. However, though astronauts such as John Young made a number of leaps while on the moon, they never jumped more than about 18 inches in height, while they were *theoretically capable of slow backflips* !

Brian believes that this is explained by the evidence that the moon's gravity is not one sixth, but much heavier, and that the spacesuits did not weigh as much as NASA said. Young and other astronauts were, essentially not capable of jumping more than 18 inches off the ground. Evidence that the suits weighed far less than 185 pounds is given that the astronauts practiced in the same suits at an area of Oregon known as the "Bend". What would be the point of doing maneuvers in a 185 pound space suit, when on the moon the astronaut and suit would weigh a mere combined total of 62 pounds? Furthermore, how did they manage to do maneuvers if the suits weighed so much in the first place? Brian estimates from his own researches that the spacesuits and backpacks weighed a combined total of about 75 pounds.

Brian also discusses the scene in which during Apollo 12 when astronaut Conrad jumped the final three feet from the bottom of the ladder to the moon's surface. He mentioned that the three foot jump may have been a short one for Neil, but for him it was a long one. Later, when he was scooping up a contingency sample of moon material, astronaut Bean warned him not to fall over since he appeared to be leaning forward too far. Supposedly, it would be difficult for him to get up in the moon suit if he fell over.

Brian analyses the scene by saying that jumping off a three foot ladder in one sixth gravity would be like jumping from a 6 inch height on the earth. Even with a heavy suit and backpack on, it would scarcely have been felt the astronaut, and they could have lowered themselves down with their arm strength alone. Furthermore, should Conrad have fallen over, they should have been able to right themselves with a simple arm push.

Furthermore, astronaut Charles Duke during Apollo 16 fell a number of times while on the moon. Since objects on the moon would supposedly take two and half times longer to fall in one-sixth gravity, Duke should have had plenty of time to catch himself.

Brian finds it incredible that the astronauts during Apollo 14 failed to climb cone crater as was planned. At one point during the mission, astronaut Shepard went down on one knee to pick up a rock and required the aid of astronaut Mitchell to stand up! About two-thirds of the way to their destination, their hearth rates were up to 120 beats per minute as the moved uphill. After four hours of travel, the two astronauts were still a half hour away, they estimated from their goal, and Shepard estimated that they could not reach the top of the crater in that time, so they abandoned the mission.

This is all the more astonishing in that the crater is little more than a hill, was the entire distance to be traveled was estimated as 1.8 miles. If the astronauts were two-thirds of the distance, they should have been able to travel the remaining half a mile in six minutes, assuming that they traveled a speed of five miles an hour on the moon in one-sixth gravity. And yet, they estimated that half an hour would not be enough!

Brian comments on a great deal more material which he finds inconsistent in the Apollo missions, including inconsistencies in the Moon Rover, but his final comment is quite reveiling.

During Apollo 17 astronauts Cernan and Schmitt began their first assignment by deploying and loading the Rover. Cernan aparently became quite excited and his Capsule Communicator, astronaut Parker warned him that his metabolic rate was going up. This meant that he was using more oxygen. Cernan replied that he had never felt calmer in his life and indicated to Parker that they would take it easy. He mentioned to Parker that he thought it was due to getting accustomed to handling himself in "Zero G."

Parker, an astronomer, then stated that he thought Cernan was working at one-sixth gravity. Cernan's reply was, "Yes. You know where we are . . .whatever." Brian suggests that the latter remark by Cernan in response to the moon's gravity seems to suggest that he wanted to avoid the discussion. Perhaps Parker was not aware of the high gravity situation and asked an embarrassing question.

Brian's book is fascinating, though not without flaw. For instance, he does not seem to understand the difference between the fundamental physical concepts of mass and weight. However, his discussion of the discrepancy of stated values of the neutral point between the earth and the moon appears valid. If Brian is correct, and the moon's gravity is nearly that of the earth's, then we are faced with the question of how and if NASA "really did it."

Perhaps NASA just faked it, as suggested in the movie **Capricorn One** and discussed in the book, **We Never Went To The Moon** . On the other hand, Brian and others suggest that we did in fact go to the moon, but we defeated the high gravity conditions not with rockets, but with anti-gravity devices.

That the Lunar Module was in reality an anti-gravity device has been suggested by several researchers. Brian claims that in high lunar gravity conditions, the LM could not have taken off from the moon, and that it's ascent is not consistent with that of a rocket powered vehicle. He also states that no rocket exhaust can be seen from the lunar module after it's initial explosion.

Does the moon have a high surface gravity, greater than one-sixth? Does gravity control really exist? If so, why do we continue to expend billions of dollars on dangerous rockets that can explode on take-off? If NASA has developed anti-gravity devices based on Einstein's unified field, then have we really stopped going to moon? Perhaps the science-fiction of the Star Wars movies is more of a reality than NASA would have us believe!

TALES FROM THE RED PLANET: MARS, UFOs, AND ANTI-GRAVITY

BY DAVID HATCHER CHILDRESS

> The improver of natural knowlege absolutely refuses to acknowledge authority as such; for every great advance in natural knowledge has involved the absolute rejection of authority.
>
> T. H. Huxley

As if the moon were not the source of enough mysteries and cover-ups by NASA and the scientific community, in their apparent attempt to hide their own "flying saucer" research and other military schemes, the Red Planet, Mars, adds its two bits to the whole bizarre scenerio.

Mars has featured in mankind's fantasies and mythology for thousands of years. The planet itself is named after the Roman god of war (or vice versa). Jonathan Swift wrote in "Gulliver's Travels" in 1726 that astronomers on "Laputa," the mythical floating land which also means "prostitute" in Spanish, had discovered two swiftly moving moons on Mars, and provided information on their distances from Mars and their periods of revolution about Mars. Astonishingly, the moons of Mars had not been "discovered" yet, and would not be, officially for another hundred and fifty years or so, though Kepler had surmised before Swift's time that Mars had two moons. Swift's information on the distances from the planet and the revolution periods of Mars were two moons. Most Scientists merely pass it off as a "good guess."

This was just the start of the mystery of the moons of Mars, and of Mars itself. Prior to 1877, when the moons were seen for the first time, no one (except possibly Swift) had seen any moons on Mars, even though excellent telescopes were at the disposal of astronomers, easily capable of discerning the moons. Mars was a popular planet to view, and literally hundreds of astronomers observed Mars for some time, and even discovered moons on Uranus until one day Asaph Hall found that Mars had two satellites where none had been before.

Not only this, but the two Martian satellites moved at very high speeds and, strangely, travelled in different directions! These and other factors have led some prominent astronomers to actually put forth the supposition that the moons are artificial!

The Soviet astronomer I.S. Schklovsky, pointed out that the Martian satellite Phobos exhibits a strange acceleration in its orbit, an irregularity which would be expected if the satellite were in reality a huge metallic sphere that was hollow. The same difference in speed, however, would be impossible for a natural astronomical body. Therefore, says Dr. Schklovsky, at least one of the moons of Mars is not a natural object, but an artificial satellite placed in orbit around the planet, possibly in 1877, or shortly before that time.

A few years later, astronomers observing Mars began to notice markings on the planet that seemed to be connected in a system that covered the whole planet. These markings were dubbed to be "canals" and were actually thought to be just that by many of the foremost astronomers of the time. The Italian astronomer Schiaparelli was the first to note the canals in the early 1890s, and other astronomers began to notice them as well. The American astronomer Percival Lowell, who built one of the best observatories in the world in Flagstaff, Arizona, became obsessed with the canals, drew detailed maps of them and worried that the Martians were fighting a losing battle to survive on a dying planet.

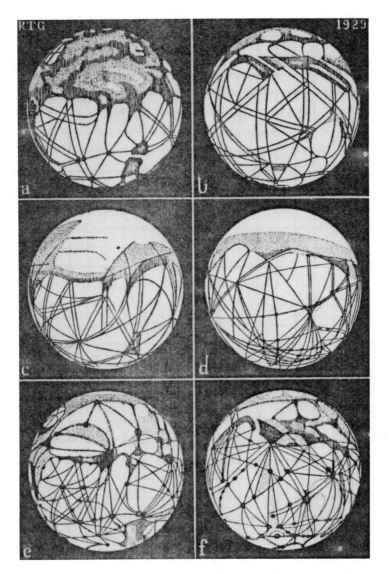

The Canals of Mars, as mapped by three of their most prominent exponents

a, b, after Schiaparelli   c, d, after Lowell   e, f, after Maggini

NOTE.—a, c, and e show exactly the same aspect of Mars (merid. longitude 0°): so do b, d, and f (merid. longitude 180°). It will be noted that the members of each triplet differ very widely in detail.

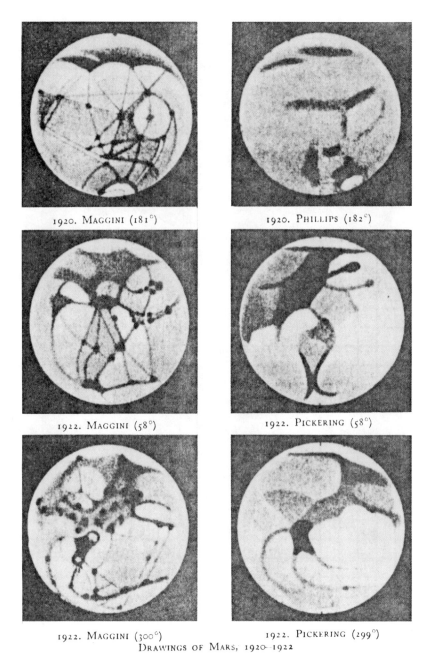

1920. MAGGINI (181°) 1920. PHILLIPS (182°)

1922. MAGGINI (58°) 1922. PICKERING (58°)

1922. MAGGINI (300°) 1922. PICKERING (299°)

DRAWINGS OF MARS, 1920–1922

By Dr. M. Maggini, the Rev. T. E. R. Phillips, and Dr. W. H. Pickering

The figures in brackets show the meridian longitude of the centre of the disc, which, it will be noticed, is practically the same for each horizontal pair

The 1922 Maggini drawings were made with slightly higher powers, and better conditions of visibility, than the remaining four

Reproduced by courtesy of Dr. W. H. Pickering

The "canals" exist, there can be no doubt. Just what they are, is the question. Considering the great variety of canals on the Martian surface, it was thought by some scientists that the Martians were trying to signal us. At one time plans were suggested for planting mid-western crops in patterns by way of acknowledging the communication. Canals were seen to wax and wane by astronomers, and would apparently move at times, confusing everybody.

In the 1910 issue of "Nature," the astronomer James Worthington made the comment, after visiting Lowell at his observatory in Flagstaff (and Lowell was of the outspoken opinion that there was life on Mars and the canals were of an artificial nature), "As to the deductions which Dr. Lowell had drawn from his observations I have nothing to say except that the startlingly artificial and geometrical appearance of the markings did force itself upon me."

Flashes of light were frequently seen on Mars and have been called Transient Martian Phenomena, much like the Transient Lunar Phenomena of the same nature. While some astronomers interpreted it as signals, others thought it to be clouds drifting across the surface. One prominent place where the "projections" (flashes of light) occurred, is the Icarium Mare, and Percival Lowell stated at the American Philosophical Society meeting in December 1901, that the more than 400 projections seen in the Mare were clouds reflecting light. Icarium Mare, he said, was undoubtedly a great tract of vegetation and was given to forming clouds.

As time went on, and Percival Lowell died, other scientists were sure to make statements that there was no life on Mars, nor any of the other planets in our solar system. This did not stop the wave of hysteria when Orson Wells broadcast his Halloween hoax of H.G. Wells' "War Of The Worlds," simulating an actual invasion from Mars.

After the Viking 1 Orbiter flew by Mars on July 31, 1976 at an altitude of 1,278 miles, taking pictures of the Martian surface, some new and interesting information suddenly came up in a photograph released and described by NASA only as "the northern latitudes of Mars." In the photograph since published several times in Omni magazine (April, 1982 and March, 1985), as well as in other journals (and reprinted in entirety here), a huge rock formation that looks like a face can be seen in the center of the photo. This rock formation has been measured as one mile across. NASA claims that it is an illusion caused by the angle of the sun.

Furthermore, to the left of the photo are two rock formations which appear to be pyramidical in shape. They are clearly throwing out triangular shadows. Parallel lines, looking like perfectly straight runways or roads appear in the upper left hand portion on the photo. According to Jim Safran of Lunar Photos in Van Nuys, California, these markings appear in quite a few of the Viking Mars photos. Oddly, these artificial-looking markings have been cropped out of the photos that appeared in Omni magazine and are not mentioned at all.

Fortunately, two computer scientists who work for Computer Science Technicolor Associates, of Seabrook, Maryland, a company that does contract work for NASA, noticed the photos, and decided to analyse them themselves. The scientists, Vincent DiPietro and Greg Molenaar concluded that the face in the photos, taken of the Elysian Plains, would "appear to have been carved rather than formed by nature," as there is no surrounding sediment that could have resulted from natural erosion, as NASA claimed.

The Face of Mars

This photo of the Martian surface was taken in July, 1976, by the Viking 1 Orbiter at a height of 1,162 miles. In its caption NASA said: "The picture shows eroded mesa-like landforms. The huge rock formation in the center, which resembles a human head, is formed by shadows giving the illusion of eyes, nose and mouth. The feature is 1.5 kilometers (one mile) across. The speckled appearance of the image is due to bit errors emphasized by enlargement of the photo."

Despite its presumably natural origin, this feature has details other than the face which give it an artificial look. At far right is an unusual rectangular formation resembling a crypt, while at left center there is an indentation as if the ground had settled around a circular structure. At top left there is a T-shaped set of lines leading to a sharp-edged rock which is topped by a noticeable black mark or opening. The whole area is suggestive of a special place of assembly.

NASA.

They furthermore concluded that the face was truly symmetrical (they were using computer enhanced photos to make their detailed analysis). It had two halves, each containing an "eye," a "cheek" and an appropriate continuation of the "mouth." They even discovered what resembles an eyewall with a visible pupil in the eye socket. They also discovered that there was a second NASA photograph of the "Face in Space," as it has been dubbed, and that it was just an illusion caused by the angle of the sun on a natural formation.

The whole affair got even stickier when a science writer who was a friend of DiPietro and Molenaar, named Richard Hoagland, got original copies of the photos and then claimed to have found, in the same photo, a "lost civilization on Mars" (Omni, Vol. 7 No. 6, March, 1985)! They had turned their attentions to the pyramidical features to the west of the face, and to the grid-like pattern of rectilinear markings like the layout of a city in the shadows of the upper pyramid. He also spotted a series of right angles contributing to an overall impression of a main avenue leading toward the face.

Hoagland discovered that this main "avenue" seems to be aligned in a special way with the face, which itself runs along a northeast-southeast axis with the Martian poles. Back then, a person standing in the center of the "city," gazing east over the face, would be sighting along a solstice alignment; that is, seeing the sun rise directly over the face on the longest day of the Martian year. Hoagland surmised that for 50,000 years, the first summer sun of the year would have risen above the face. Later, as the planet tilted, the alignment of the solstice viewing would have passed right through the top of the pyramid as well.

The honeycomblike tracery that exists in the shadow of the pyramid could be optical "glitches" caused by the photo-enhancing process. Yet, these walls cast shadows, and DiPietro and Molenaar claim they did not get these kinds of glitches with enhancements of aerial photos taken here on Earth. The grid spacing suspiciously resembles that of real city streets, and the layout is aligned toward the winter solstice sunrise. An architect friend of Hoagland's calculated the buildings would have been oriented in a manner that would best use the scant winter warmth of the shortest day of the Martian year.

Incredibly, the senior scientist with the U.S. Geological Survey and one of the world's leading experts on Martian geology, a man who headed the NASA group in charge of selecting the sites where the Viking landers set down, Harold Masursky, told Omni magazine, "If you're going to say features like that are evidence for a past civilization, that's total nonsense. I'm working on finding landing sites for a possible Mars Rover. And this (the city on Mars) is not one of the areas where I would send what is probably a thirty-billion dollar mission. In fact, if somebody bought us a free one, I'm not sure I'd send it there because there are too many other places that are more interesting."

Is Masursky a blind idiot, or is he toeing the party line, so-to-speak? He is probably not an idiot, and considering the startling finds of NASA on the Moon, and their many secret projects and cover-ups, it is not surprising that he would make such statements. Either masursky is on the periphery of NASA knowledge, and actually believes that there is nothing worth viewing at this "Martian City" (it would be interesting to hear his comments on UFOs), or he is naturally trying to avert attention from this startling find and cover up what may even be a "live city" as opposed to a "dead" one.

Interestingly, a book published in 1978 by Avon books in the United States, and which originally appeared as a BBC special in Britain, called "Alternative 3" (by Leslie Watkins, David Ambrose and Christopher Miles) was reportedly an investigation into the disappearance of scientists in Britain and the United States. According to the book, these scientists were being sent to Mars (!) by NASA to work in secret cities there, in an effort to create a habitual climate on Mars (which included melting the polar ice caps and building dome-cities on the planet). The reason, the book stated, was that NASA was doing this because the Earth's atmosphere is becoming super-heated and unbreathable.

Three photographers in different locations obtained shots of this low-flying vehicle that appeared over the countryside near Madrid, Spain in January, 1967 (an 'invasion' year). The object was estimated at 50 to 80 feet in diameter. The emblem is comparable to the astrological sign of Uranus. These photos first appeared in the Spanish publication Horizonte, directed by Antonio Ribera.

Life was doomed on earth; therefore, a secret conspiracy, involving most of the world's governments (including the U.S.S.R.) were working to move a certain portion of mankind to Mars.

According to the book, the first manned landing on Mars took place in the early 1960s, and Mars Bases were begun shortly afterwards. Anti-gravity spaceships were used to shuttle scientists and "brainwashed," kidnapped, slave-workers to the bases to work. There was a lunar staging base in a crater on the moon.

While the book made some impact on UFO and conspiracy buffs, and was written in a matter-of-fact style and purported to be an investigative book, it was in fact an April Fool's Day television special done for the BBC that was never aired on April Fool's Day because of a television strike in Britain. When it was finally aired, straight-faced, some months later, most people did not realize that it was an April Fool's joke. It was later published in Britain and the United States in book form as "science fact," also on April Fool's Day. An interesting story, it is however quite unlikely, considering the difficulty that both the Soviets and the Americans had in just landing space probes on the moon in the early sixties, much less Mars. Even if NASA possessed anti-gravity vehicles at that time, it seems unlikely that they would have been flying them to Mars while laughingly flinging Rangers at the Moon. Furthermore, the author, British journalist Leslie Watkins has come right out and said that the whole book was a hoax. Yet still, many UFO buffs believe it as fact.

One does wonder, though, if NASA has visited Mars yet in one of its supposed "Anti-Gravity" craft? It seems unlikely. They are far too busy preparing for nuclear war and setting up presumed bases on the Moon. With all the UFOs out there (especially on the Moon, if reports are to be believed), and their occasional interference in the NASA space program, NASA may consider it too dangerous to take a manned flight to Mars.

One cannot help but think that there was certainly life on Mars in the past, if not now. Are some UFOs from Mars? Perhaps they are little green men, busy mining our Moon without a permit, exploiting Earth's God-given natural resources, much as huge multi-national companies and western powers have been doing to underdeveloped nations here on earth.

Still more disturbing are the many reports of captured alien craft and even aliens, by the U.S. Government. In the book, Clear Intent: The Government Coverup Of the UFO Experience, by Lawrence Fawcett and Barry J. Greenwood, they catalogue a great deal of evidence for what they term a massive coverup of UFO data and even captured UFOs that have either crash landed or been shot down by the airforce.

They relate one interesting story about a "crashed" UFO in the Pacific that happened in 1973. An unidentified Naval Intelligence officer tells how, while stationed at the Great Lakes Naval Base in Chicago, he was the Officer of the Guard and was asked to take a sealed message to the Commander inside a highly restricted quonset hut at the northwestern end of the base. He had been told that there was highly top secret material inside. Normally the OD would come to the door, but was busy that night so he was allowed inside. This was highly unusual.

"As I went to the doorway, where the OD was, I saw a very highly unusual craft over to my left. The craft was possibly thirty to thirty-five feet long, about twelve to fifteen feet at its thickest part; then it tapered off in the front to a teardrop shape. I only caught it at an angular view. It looked like it did not have any seams to it. It had a bluish tint, but that was only if you looked at it for a few seconds."

As the officer turned to leave, he got another look at it, "At this time I had a very good view about halfway from the craft to the tail section. The whole craft tapered back to a very high edge. It looked as if it had a razor edge, a razor sharp edge. The bottom went about three quarters the length of the craft and then angled sharply upward." The craft sat on a frame made out of four by four wooden

San Francisco Chronicle Monday, September 16, 1985

U.S. Contesting Lawsuit Over UFO Radiation

Houston

Three people suing the federal government for $20 million say they do not know to this day what it was that hovered far over their heads and zapped them with radiation almost five years ago.

They claim it was an unidentified flying object that was escorted away by military helicopters. They say they suspect it was a secret U.S. military experiment. In any event, they say, the government should have warned residents that a UFO was in their area.

The military says it had nothing to do with the alleged occurrence Dec. 29, 1980, on a rural road northeast of Houston. Even if there were UFOs, a U.S. attorney says, the government has no duty to warn people about them because the government does not know whether they are dangerous.

With such straightforward arguments, the government is urging U.S. District Judge Ross Sterling to throw out a lawsuit filed here last year by former Dayton, Texas, cafe owner Betty Cash, 56, former waitress Vickie Landrum, 61, and Landrum's grandson, Colby Landrum, 11.

The suit claims that about 9 p.m. on Dec. 29, 1980, while headed for their homes in Dayton along a two-lane road about 30 miles northeast of Houston, the three encountered a brightly glowing craft the size of a city water tower. It hovered at treetop level, had red and orange flames flowing from its bottom, and bathed them in intense heat for several minutes before it was escorted away by at least 23 helicopters, they assert.

Their lawsuit, filed under the Federal Tort Claims Act, claims that the government failed to warn them of the UFO and "negligently, carelessly and recklessly" allowed it "to fly over a publicly used road and come in contact with the plaintiffs."

As a result, all suffered stomach pains, vomiting, diarrhea, radiation burns, deteriorating eyesight, and the women's hair fell out and grew back with a different texture, the suit claims. It also says they became highly sensitive to sunlight, suffered blisters and headaches, and that Cash developed breast cancer.

Despite the plaintiffs' report that there were no markings on the "large, unconventional aerial object" or on the helicopters, the lawsuit has recently moved from simply suggesting that the government merely knew about the UFO to saying that the government owned the UFO.

The government has offered affidavits from high-ranking officials that it had nothing to do with the UFO.

Dallas Times Herald

blocks, with crossbeams under it, so that it was sitting one or two feet off the floor.

Later, the Intelligence Officer was in San Diego talking to some crew members of a destroyer who said that they had tangled with an unidentified craft. The destroyer had shot down the craft while heading from San Diego to Hawaii in 1973 with a surface-to-air missle, but did not destroy it. It sank in about 350 feet of water. The Glomar Explorer was used to extract the craft and it was sent by rail to Chicago. When the crew member drew a picture of the unidentified craft, it matched perfectly the craft that the Intelligence Officer had seen at the Great Lakes Naval Base.

One interesting incident reported in 1981 by many newspapers as well as OMNI and other magazines, was the case of three Texas school teachers who saw a "flaming" flying saucer cruise over their car in the desert and crash beyond a hill. Just behind this "flying saucer" was an Army helicopter, which landed at the crash site. The ladies, all middle-aged, and respectable, reported their sighting to the police. Then, a few days later, all three of the women lost all the hair on their heads! They had apparently gotten a dose of radiation!

This case does not appear to be one of an alien ship, but rather an unsuccessful test of an experimental U.S. military craft. Other tales of crashed UFOs appear in many books; the most famous is probably "the Roswell Incident," detailed by Charles Berlitz in the book, "The Roswell Incident" (see bibliography). While many of the stories appear rather fanciful in nature, and most "kidnapping" stories appear to be utter bunk, one can not discount the possibility that the U.S. and other governments have captured a UFO.

Front page of a Soviet 'popular sicence' type magazine, purporting to show a fantastic jet-propelled disc type passenger plane then (1964) being contemplated by Soviet engineers. It never was heard of again.

The above photocopy (unfortunately of poor quality) purportedly shows a captured alien in the hands of the American military. (Note the computer digitized version below). On May 22, 1950, an unnamed informant turned the original of this photograph over to agent John Quinn of the New Orleans FBI Field Office claiming he had purchased the photograph from another individual for the sum of $1.00 and was "placing it in the hands of the goverment" because it pictured "a man from Mars in the United States". The picture, supposedly showing an alien survivor of a UFO crash in the custody of two U.S. military policemen, reportedly first surfaced in Wiesbaden, Germany, in the late 1940s allegedly in the possession of a U.S. GI stationed there at the time. How he came into possession of such a picture remains unclear, as do the identities of the two soldiers portrayed, the location of the military base and the nature of the portable respiratory apparatus that is obviously being used to assist the alien's breathing. Receiving some limited publicity in Germany during the late 40s, the photo and story was naturally regarded with skepticism by U.S. officials.

Dr. Zitzenpop's Generic Anti-Gravity Equation

$$G = \frac{STP \approxeq \wp \lesseqgtr Q \pm K_{\ddot{W}} [\text{Theory Z}] \therefore ?}{\pm\infty \sqrt{ie^{\underline{2}}}}$$

Legend for Dr. Zitzenpop's Monumental Work:

- **G** is the unit of gravitational force (one earth G is equal to 1×10^{-6} gollygees)
- **STP** is the unit of lubrication
- **$\wp$** is a shep, equal to 1.37 moecurlys
- **Q** is the charge in coulombs
- **K** is the dielectric constant
- **i** is an imaginary number, just make one up.
- **e** is equal to the number of virtual particles that make up an electron divided by the absolute value of pi.
- **$\ddot{W}$** a dimensionless constant. Ignore it— maybe it'll go away
- **Theory Z** a Japanese constant
- **$\approxeq$** probably has nothing to do with
- **$\lesseqgtr$** greater than, less than, or perhaps equal to
- **$\pm$** plus or minus an unknown sum
- **?** self explanatory
- **$\underline{2}$** I'll let you know when I look it up
- **$\therefore$** therefore
- **$\sqrt{}$** square root of
- **$\pm\infty$** plus or minus an infinite sum

Let us now scrutinize the work of the late towering genius of our time, Dr. Lazlo Zitzenpop. His solution to the antigravity problem is... let us say... unique. He recently published his unified field theory in the back of a technical magazine. Actually it was in a 2" by 2" ad. But quantity in this case is not important--- it is quality that counts. Getting into the equation we see the unit STP. This actually stands for a particle that he once discovered called Space Temporal Precessionals. This particle is the antiparticle of gluons and makes a particle slippery. The short form of this is called STePons. They give the same effect to the particle as you or I walking on ball bearings. They also change a quarks' color to black and blue. Dr. Zitzenpop thought that it didn't have any bearing on the problem but he put it in anyway. The next unit is the shep. This is the unit of credibility in a particular relation, but I don't believe in it so let's move on. The next unit is Q, the charge in coulombs (not to be confused with another unit, the hair-coulomb). And the next unit is K, the dielectric constant which is a particular materials' ability to store an electric field. All insulators have a particular dielectric constant rating except for certain cereals which have their own special K. Of course the next unit is i which is an imaginary number. Just make one up. The next one is e. This particular unit is one wrapped in esoteric meaning. It is eqivalent to the number of virtual particles that make up an electron divided by the absolute value of pi, both of which express changing infinite values. What it really expresses is infinity divided by infinity which reduces down to 1: a really elaborate method of describing the change of unity over time. The next unit is really esoteric. It is a dimensionless constant which also lacks a definition as well as a function. I really think the doctor used it to get the spacing right. The next one is Theory Z, which is a Japanese constant. It is also the unit of ambition of the quark (which is also the sound made by a durk). The operators have been fully explained as to their function on the previous page except for "?" which represents an unknown virtual operator nested in the tenth level. So much for brand x.

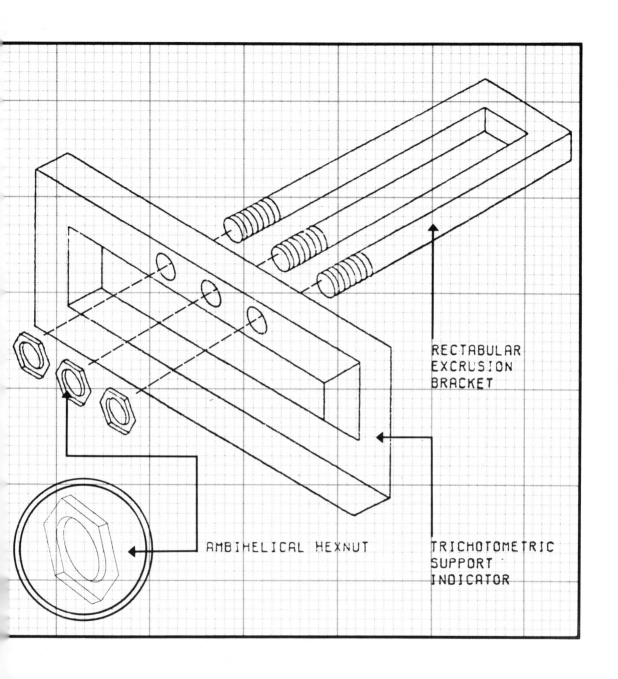

Dr. Zitzenpop's Superhelical Computer Ultraplot Machine (S.C.U.M), which was combined with his Bidimensional Axiometric Gravitometer (B.A.G.), came up with this indispensible component for his version of an Antigravity device.

PHOTO BY JOHN HANNAH/RDR

Anti-Gravity

Fashions of the 30's

Looking your best in space is easy in this stunning 3-piece jumpsuit.

This fashion concious pair are ready for an evening out on the town.
(Particle beam weapon sold separately)

Perfect for entertaining, throw on these snappy outfits when extraterrestrial guests arrive.

ANCIENT INDIAN AIRSHIP TECHNOLOGY
BY DAVID HATCHER CHILDRESS

Many researchers into the UFO enigma tend to overlook a very important fact. While it is mostly assumed that most flying saucers are of alien, or perhaps Governmental Military origin, another possible origin of UFOs is ancient India and Atlantis.

What we know about ancient Indian flying vehicles comes from ancient Indian sources; written texts that have come down to us through the centuries. There is no doubt that most of these texts are authentic; many are the well known ancient Indian Epics themselves, and there are literally hundreds of them. Most of them have not even been translated into English yet from the old sanskrit.

The Indian Emperor Ashoka started a "Secret Society of the Nine Unknown Men": great Indian scientists who were supposed to catalogue the many sciences. Ashoka kept their work secret because he was afraid that the advanced science catalogued by these men, culled from ancient Indian sources, would be used for the evil purpose of war, which Ashoka was strongly against, having been converted to Buddhism after defeating a rival army in a bloody battle.

The "Nine Unknown Men" wrote a total of nine books, presumably one each. Book number six was "The Secrets of Gravitation!" This book, known to historians, but not actually seen by them, dealt chiefly with "gravity control." It is presumably still around somewhere, kept in a secret library in India, Tibet or elsewhere (perhaps even in North America somewhere). One can certainly understand Ashoka's reasoning for wanting to keep such knowledge a secret, assuming it exists. Imagine if the Nazis had such weapons at their disposal during World War II. Ashoka was also aware of the devastating wars using such advanced vehicles and other "futuristic weapons" that had destroyed the ancient Indian "Rama Empire" several thousand years before.

Only a few years ago, the Chinese discovered some sanskrit documents in Lhasa, Tibet and sent them to the University of Chandrigarh to be translated. Dr. Ruth Reyna of the University said recently that the documents contain directions for building interstellar spaceships!

Their method of propulsion, she said, was "anti-gravitational" and was based upon a system analogous to that of "laghima," the unknown power of the ego existing in man's physiological make-up, "a centrifugal force strong enough to counteract all gravitational pull." According to Hindu Yogis, it is this "laghima" which enables a person to levitate.

Dr. Reyna said that on board these machines, which were called "Astras" by the text, the ancient Indians could have sent a detachment of men onto any planet, according to the document, which is thought to be thousands of years old. The manuscripts were also said to reveal the secret of "antima"; "the cap of invisibility" and "garima"; "how to become as heavy as a mountain of lead."

Naturally, Indian scientists did not take the texts very seriously, but then became more positive about the value of them when the Chinese announced that they were including certain parts of the data for study in their space program! This was one of the first instances of a government admitting to be researching anti-gravity.

The manuscripts did not say definitely that interplanetary travel was ever made but did mention, of all things, a planned trip to the Moon, though it is not clear whether this trip was actually carried out. However, one of the great Indian epics, the Ramayana, does have a highly detailed story in it of a trip to the moon in a Vimana (or "Astra"), and in fact details a battle on the moon with an "Asvin" (or Atlantean") airship.

This is but a small bit of recent evidence of anti-gravity and aerospace technology used by Indians. To really understand the technology, we must go much further back in time.

The so-called "Rama Empire" of Northern India and Pakistan developed at least fifteen thousand years ago on the Indian sub-continent and was a nation of many large, sophisticated cities, many of which are still to be found in the deserts of Pakistan, northern, and western India. Rama existed, apparently, parallel to the Atlantean civilization in the mid-Atlantic Ocean, and was ruled by "enlightened Priest-Kings" who governed the cities. The seven greatest capital cities of Rama were known in classical Hindu texts as "The Seven Rishi Cities."

According to ancient Indian texts, the people had flying machines which were called "Vimanas." The ancient Indian epic describes a Vimana as a double-deck, circular aircraft with portholes and a dome, much as we would imagine a flying saucer.

It flew with the "speed of the wind" and gave forth a "melodious sound." There were at least four different types of Vimanas; some saucer shaped, others like long cylinders ("cigar shaped airships"). The ancient Indian texts on Vimanas are so numerous, it would take volumes to relate what they had to say. The ancient Indians, who manufactured these ships themselves, wrote entire flight manuals on the control of the various types of Vimanas, many of which are still in existence, and some have even been translated into English.

The Samara Sutradhara is a scientific treatise dealing with every possible angle of air travel in a Vimana. There are 230 stanzas dealing with the construction, take-off, cruising for thousand of miles, normal and forced landings, and even possible collisions with birds. In 1875, the Vaimanika Sastra, a fourth century B.C. text written by Bharadvajy the Wise, using even older texts as his source, was rediscovered in a temple in India. It dealt with the operation of Vimanas and included information on the steering, precautions for long flights, protection of the airships from storms and lightening and how to switch the drive to "solar energy" from a free energy source which sounds like "anti-gravity."

The Vaimanika Sastra (or Vymaanika-Shaastra) has eight chapters with diagrams, describing three types of aircraft, including apparatuses that could neither catch on fire nor break. It also mentions 31 essential parts of these vehicles and 16 materials from which they are constructed, which absorb light and heat; for which reason they were considered suitable for the construction of Vimanas. This document has been translated into English and is available by writing the publisher: VYMAANIDA-SHAASTRA AERONAUTICS by Maharishi Bharadwaaja, translated into English and edited, printed and published by Mr. G. R. Josyer, Mysore, India, 1979 (sorry, no street address). Mr. Josyer is the director of the International Academy of Sanskrit Investigation located in Mysore.

There seems to be no doubt that Vimanas were powered by some sort of "anti-gravity." Vimanas took off vertically, and were capable of hovering in the sky, like a modern helicopter or dirigible. Bharadvajy the Wise refers to no less than 70 authorities and 10 experts of air travel in antiquity. These sources are now lost.

Vimanas were kept in a Vimana Griha, a kind of hanger, and were sometimes said to be propelled by a yellowish-white liquid, and sometimes by some sort of mercury compound, though writers seem confused in this matter. It is most likely that the later writers on Vimanas, wrote as observers and from earlier texts, and were undestandably confused on the princicple of their propulsion. The "yellowish-white liquid" sounds suspiciously like gasoline, and perhaps Vimanas had a number of different propulsion sources, including combustion engines and even "pulse-jet" engines. It is interesting to note, that the Nazis developed the first practical pulse-jet engines for their V-8 rocket "buzz bombs." Hitler and the Nazi staff were exceptionally interested in ancient India and Tibet and sent expeditions to both these places yearly, starting in the 30's, in order to gather esoteric evidence that they did so, and perhaps it was from these people that the Nazis gained some of their scientific information!

According to the Dronaparva, part of the Mahabarata, and the Ramayana, one Vimana described was shaped like a sphere and born along at great speed on a mighty wind generated by mercury. It moved like a UFO, going up, down, backwards and forewards as the pilot desired. In another Indian source, the Samar, Vimanas were "iron machines, well-knit and smooth, with a charge of mercury that shot out of the back in the form of a roaring flame." Another work called the Samaranganasutradhara describes how the vehicles were constructed. It is possible that mercury did have something to do with the propulsion, or more possibly, with the guidance system. Curiously, Soviet scientists have discovered what they call "age-old instruments used in navigating cosmic vehicles" in caves in Turkestan and the Gobi Desert. The "devices" are hemispherical objects of glass or porcelain, ending in a cone with a drop of mercury inside.

It is evident that ancient Indians flew around in these vehicles, all over Asia, to Atlantis presumably; and even, apparently, to South America. Writing found at Mohenjodaro in Pakistan (presumed to be one of the "Seven Rishi Cities of the Rama Empire") and still undeciphered, has also been found in one other place in the world: Easter Island! Writing on Easter Island, called Rongo-Rongo writing, is also undeciphered, and is uncannily similar to the Mohenjodaro script. Was Easter Island an air base for the Rama Empire's Vimana route? (At the Mohenjo-Daro Vimana-drome, as the passenger walks down the concourse, he hears the sweet, melodic sound of the announcer over the loudspeaker, "Rama Airways flight number seven for Bali, Easter Island, Nazca, and Atlantis is now ready for boarding. Passengers please proceed to gate number...")

In the Ramayana, the Celestial Car of Rama takes Rama, the Hero, from Ceylon to Mount Kailas in Tibet, no small distance, and speaks of the "fiery chariot" thusly: "Bhima flew along in his car, resplendent as the sun and loud as thunder... The flying chariot shone like a flame in the night sky of summer ... It swept by like a comet... It was as if two suns were shining. Then the chariot rose up and all the heaven brightened."

In the Mahavira of Bhavabhuti, a Jain text of the eighth century culled from older texts and traditions, we read: "An aerial chariot, the Pushpaka, conveys many people to the capital of Ayodhya. The sky is full of stupendous flying-machines, dark as night, but picked out by lights with a yellowish glare."

The Vedas, ancient Hindu poems, thought to be the oldest of all the Indian texts, describe Vimanas of various shapes and sizes: the "ahnihotra-vimana" with two engines, the "elephant-vimana" with more engines, and other types named after the kingfisher, ibis and other animals.

Unfortunately, Vimanas, like most scientific discoveries, were ultimately used for war. Atlanteans used their flying machines, "Vailixi," a similar type of aircraft, to literally try and subjugate the world, it would seem, if Indian texts are to be believed. The Atlanteans, known as "Asvins" in the Indian writings, were apparently even more advanced technologically than the Indians, and certainly of a more war-like temperment. Although no ancient texts on Atlantean Vailixi are known to exist, some information has come down through esoteric, "occult" sources which describe their flying machines. Similar, if not identical to Vimanas, Vailixi were generally "cigar shaped" and had the capability of manuvering underwater as well as in the atmosphere or even outer space. Other vehicles, like Vimanas, were saucer shaped, and could apparently also be submerged.

According to Eklal Kueshana, author of "The Ultimate Frontier," in an article he wrote in 1966, Vailixi were first developed in Atlantis 20,000 years ago, and the most common ones are "saucer-shaped of generally trapezoidal cross-section with three hemispherical engine pods on the underside." "They use a mechanical antigravity device driven by engines developing approximately 80,000

Above left. Prehistoric Swiss rock painting of space-man.

Above right. Primitive rock painting of astronaut, Uzbekistan, USSR.

Right. A *dogu*—Japanese clay figurine of the Jomon period. The detailed resemblance to an astronaut's equipment may not be coincidental?

horse power."

The Ramayana, Mahabarata and other texts speak of the hideous war that took place, some ten or twelve thousand years ago between Atlantis and Rama using weapons of destruction that could not be imagined by readers until the second half of this century.

The ancient Mahabharata, one of the sources on Vimanas, goes on to tell the awesome destructiveness of the war: "...(the weapon was) a single projectile charged with all the power of the Universe. An incandescent column of smoke and flame As bright as the thousand suns rose in all its splendor...
An iron thunderbolt,
A gigantic messenger of death,
Which reduced to ashes
The entire race of the Vrishnis
And the Andhakas.
...the corpses were so burned
As to be unrecognizable.
The hair and nails fell out;
Pottery broke without apparent cause,
And the birds turned white.
...After a few hours
All foodstuffs were infected...
...to escape from this fire
The soldiers threw themselves in streams
To wash themselves and their equipment..."

It would seem that the Mahabharata is describing an atomic war! References like this one are not isolated; but battles, using a fantastic array of weapons and aerial vehicles are common in all the epic Indian books. One even describes a Vimana-Vailix battle on the Moon! The above section very accurately describes what an atomic explosion would look like and the effects of the radioactivity on the population. Jumping into water is the only respite.

When the Rishi City of Mohenjodaro was excavated by archeologists in the last century, they found skeletons just lying in the streets, some of them holding hands, as if some great doom had suddenly overtaken them. These skeletons are among the most radioactive ever found, on a par with those found at Hiroshima and Nagasaki. Ancient cities whose brick and stone walls have literally been vitrified, that is--fused together, can be found in India, Ireland, Scotland, France, Turkey and other places. There is no logical explanation for the vitrification of stone forts and cities, except from an atomic blast. Futhermore, at Mohenjo-Daro, a well planned city laid on a grid, with a plumbing system superior to those used in Pakistan and India today, the streets were littered with "black lumps of glass." These globs of glass were discovered to be clay pots that had melted under intense heat!

With the cataclysmic sinking of Atlantis and the wiping out of Rama with atomic weapons, the world collapsed into a "stone age" of sorts, and modern history picks up a few thousand years later. Yet, it would seem that not all the Vimanas and Vailixi of Rama and Atlantis were gone. Built to last for thousands of of years, many of them would still be in use, as evidenced by Ashoka's "Nine Unknown Men" and the Lhasa manuscript.

That secret societies or "Brotherhoods" of exceptional, "enlightened" human beings would have preserved these inventions and the knowledge of science, history, etc., does not seem surprising. Many well known historical personages including Jesus, Buddah, Lao Tzu, Confucious, Krishna, Zoroaster, Mahavira, Quetzalcoatl, Akhenaton, Moses, and more recent inventors and of course many other people who will probably remain anonomous, were probably members of such a secret organization.

It is interesting to note that when Alexander the Great invaded India more than two thousand years ago, his historians chronicled that at one point they were attacked by "flying, fiery shields" that dove at his army and frightened the cavalry. These "flying saucers" did not use any atomic bombs or beam weapons on Alexander's army however, perhaps out of benevolence, and Alexander went on to conquer India.

It has been suggested by many writers that these "Brotherhoods" keep some of their Vimanas and Vailixi in secret caverns in Tibet or some other place is Central Asia, and the Lop Nor Desert in western China is known to be the center of a great UFO mystery. Perhaps it is here that many of the airships are still kept, in underground bases much as the Americans, British and Soviets have built around the world in the past few decades.

Still, not all UFO activity can be accounted for by old Vimanas making trips to the Moon for some reason. Undoubtedly, some are from the Military Governments of the world, and possibly even from other planets. Of course, many UFO sightings are "swamp, gas, clouds, hoaxes, and hallucinations, while there is considerable evidence that many UFO sightings, especially "kidnappings" and the like, are the result of what is generally called "telepathic hypnosis." One common thread that often runs between "Alien kidnappings," "sex with aliens," and other "close encounters of a third kind" is a buzzing in the ears just before the encounter. According to many well informed people, this is a sure sign of telepathic hypnosis."

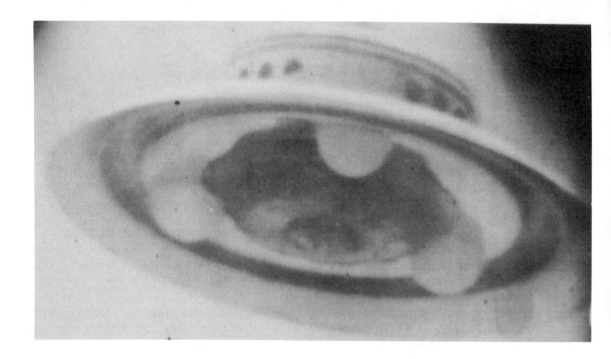

The above famous photo was allegedly taken by George Adamski in the early '50s. Among other things Adamski claims to have gone to the moon and to have traveled from Kansas City to Davenport, Iowa in this ship. It has been noted that Adamski's saucer looks strikingly like a chicken brooder shown in the lower drawing. It was also assumed that Adamski had photographed bottle cooler lids that had exactly the same design as Adamski's saucers. However, it was discovered that the bottle cooler's designer had designed the bottle cooler lid in 1959 six years after Adamski published his photographs. In fact, it turned out the designer was a follower of the UFO mystery and his design was a tribute to George Adamski! It's also worth noting that Adamski claimed that the hemispherical landing pods located on the underside of the craft were extendable landing gear. Assuming that the photograph is genuine, those pods would more likely be guidance and field stabilization components (vortex pods). Photographs courtesy of Openhead Press, London, UK.

BOTTLE COOLER LID

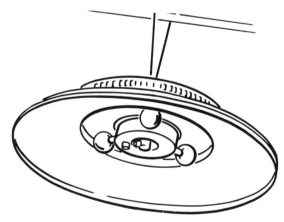

CHICKEN BROODER

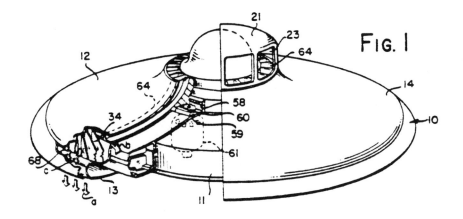

FIG. 1

A Selection of Anti-Gravity Patents

Impossible? Ce mot n'est pas francais!
(Impossible? This word is not French!)
                              -Napoleon

Antigravity Space Drives

Antigravity Drive	Researcher	Date
Coanda Effect lifting Devices	Coanda	1939
Entropy Engine	Jones	1967
Magnus Effect Lifting Devices	Swanson	1961
Non-Simultaneity Surge Drive	Davis	1962
Skyhook Via Satellite Elongation	Isaacs, Vine, Bradner & Bachus	1966
Kinetic Diode	Jones	1969
Momentum Exchanger Principle	Cox	1969
Propulsion System	Trupp	1965
Project Orion	Dyson	1968
Interplanetary Cables or Roads	Unknown	----
Full-Wave Alternating Force Rectifier By Vector Inversion	Cox	1967
Radial Force Generator	Keeney	1970
Directional Force From Rotary Motion	Nowlin	1944
Centrifugal Variable Thrust Mechanism	Laskowitz	1934
Propulsion Mechanism	Quisling	1930
Direct Push Unit	Llamorzas	1953
Momentor	Ecklin	1966
Gamma Drive	Pittman	1961
Dean System Space Drive	Dean	1959
Gyrothrust	Kellogg	1967
Flywheel Drive	Cox	1967
Inertial Propulsion System	Farrall	1966
Centrifugal Space Drive	Laurizan	1967
Rotary Energy Conversion System	Schoonrok	1969
Orbiting Mass Engine	Cox	1970
Hallberg Device	Hallberg	1966
German Copper Drive	Schauberger	1940
Internal Reaction Engines	Goodykoontx	1967
Impulse/Impact Drive	Bull	1934
Fluid Self-Moving Mechanism	Epstein	1970
Prime Mover	Auweele	1970
Directional Force Generator	Young	1971

Apparatus For Imparting Motion To A Body	Bella	1968
Propulsion Apparatus	Matyas	1971
Fluid Mercury Drive	Cox and others	1971
Porter Propulsion Systems	Porter	1972
Electrokinetic Device	Brown	1960
Levity Disk (Project Ezekial)	Searl	1968
Circular Foil OTC-X1	Carr	1950
Rotating Magnetic-Antigravity Device	Schoonrok	1968
Gravitational Machines	Dyson	1962
Protational Field Drive	Forward	1961
Etheric Vortex Drive	Roos	1970
Space Warp Drives	Henderson	1971
Magnetic-Etheric Screw	Cox	1967

Don't keep forever on the public road, going only where others have gone. Leave the beaten track occasionally and dive into the woods. You will be certain to find something you have never seen before...

—Alexander Graham Bell

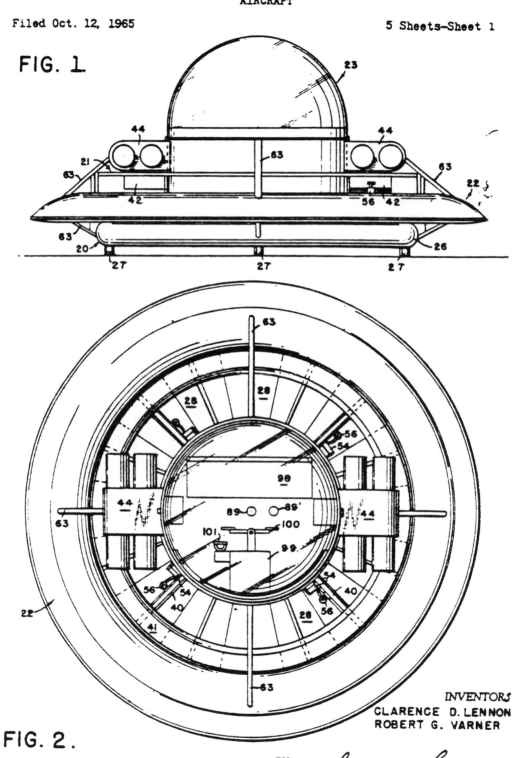

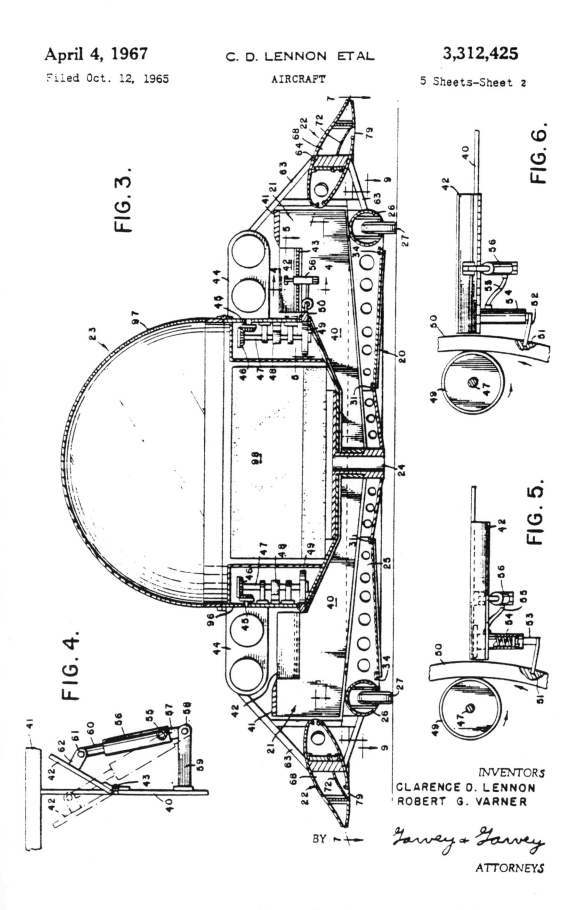

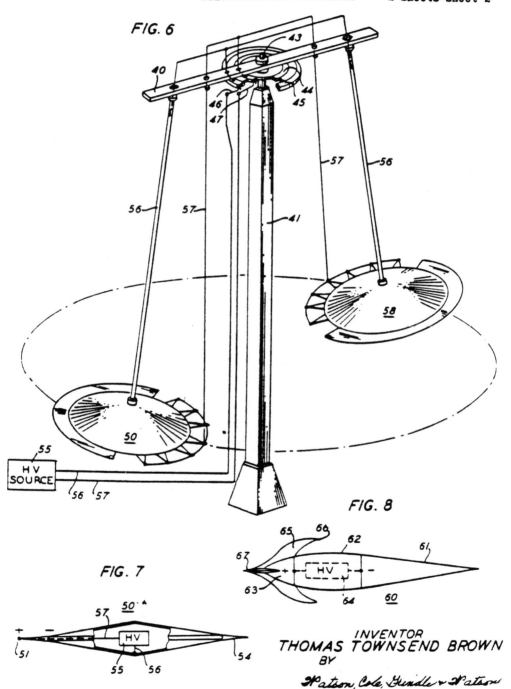

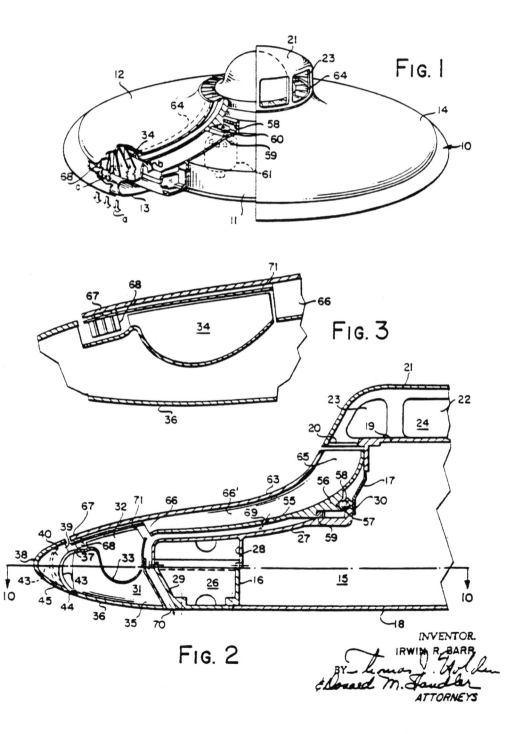

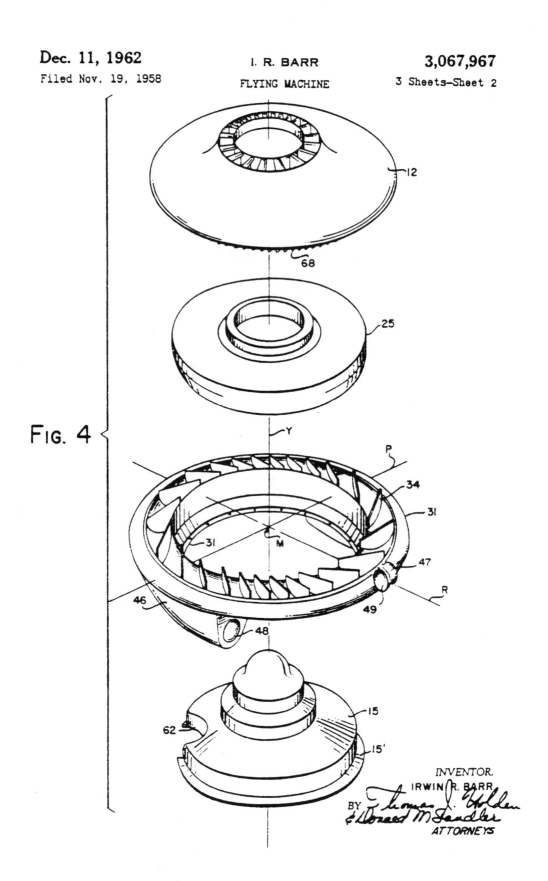

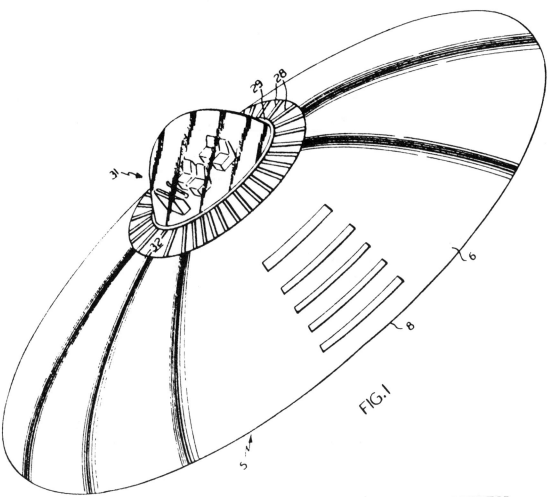

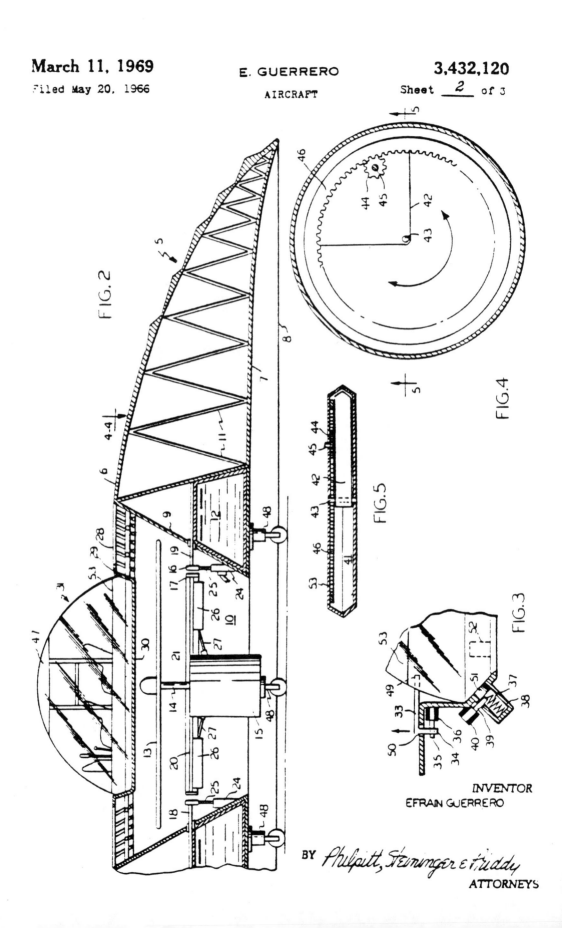

ZEPPELINS, UFO'S, ANTI-GRAVITY, AND THE MYSTERIES OF THE AIRSHIP AGE
BY DAVID HATCHER CHILDRESS

Unidentified Flying Objects (UFOs) have been sighted, according to ancient records, for thousands of years. Many sacred books have writings in them that today would undoubtedly be catagorized as UFO sightings. Among these books are the Bible, and the sightings of Ezekial have been the subject of several books, including one by two NASA employees called The Spaceships of Ezekial. Ancient sightings of UFOs are far too numerous to go into here, but a certain rash of UFO sightings in the late nineteenth century are of particular interest.

In 1873 at Bonham, Texas, workers in a cotton field suddenly saw a shiny, silver object that came streaking down from the sky at them. Terrified, thay ran away, while the "great silvery serpent"as some people described it, swung around and dived at them again. A team of horses ran away and the driver was thrown beneath the wheels of the wagon and killed. A few hours later that same day in Fort Riley, Kansas, a similar "airship" swooped down out of the skies at a cavalry parade and terrorized the horses to such an extent that the cavalry drill ended in a tumult.

In 1882, an "airship", described variously as a "cigar", "torpedo", "spindle" or "shuttle" was seen on November 17 by the Royal Observatory in Greenwich, England by the famous British astronomer E.W. Maunder and a number of his colleagues. This airship appeared quite suddenly and moved steadily across the sky telescope. Comparing notes, they discovered that they had all seen the same thing and that it was not a cloud or meteor, but an airship of some sort that was decidedly "extraordinary and alarming".

What is generally known as the "Great Airship Flap of 1897", actually began on November 22, 1896 in San Francisco. Thousands of folks going home from work witnessed a large dark "cigar" shaped object with "stubby wings" traveling northwest across Oakland. A few hours later reports came from other northern California cities; Chico, Santa Rosa, Sacramento and Red Bluff, all reporting the same thing and describing a similar airship. It is quite possible that the vehicle was heading for Mount Shasta in Northern California.

It returned a little over a week later, moving steadily against the wind, something which confounded those who had earlier insisted that it was a balloon. Throughout December, 1886, and January, February, March, and April, 1897, a rash of "silvery cigar-shaped airships" were reported across the United States, especially in the midwest. It was reported in Colorado, Kansas,Texas, Iowa, Nebraska, Missouri, Wisconsin and Minnesota, and on April 10, 1897 thousands of people in Chicago reported the ship.

The airship flew at great height, and many persons observed it through telescopes, describing it as cigar-shaped with broad, stubby wings. At night it was said to flash with red, green and white lights.

This unprecedented and much published rash of sightings throughout the midwest began apparently on April 9, 1897 and ended on April 16. During that time literally thousands of people observed the UFO, and hundreds of newspaper articles were written on it.

This was not the end, however. On April 19, the small town of Sistersville, West Virginia was awakened during the night by the loud and frantic blowing of the siren of the local saw-mill. The citizens tumbled from their beds and out into the streets to view a great, cigar-shaped airship circling overhead. Brilliant searchlights swept through town, making the community as bright as day. All agreed that it had red, green and white lights on it, was tubular in shape, and made a humming sound. After ten to fifteen minutes of this, it suddenly switched off its searchlights and headed eastward at great speed. Church attendance in Sistersville rose dramatically after this episode!

Incredibly, the Dallas Morning News for the same day, April 19, 1897 carried a story stating that on the morning of the 17th of April an "airship" was seen over Aurora, Texas and that it struck "Judge Procter's waterwell" and exploded! The story goes on to say that the pilot's body was too mutilated to describe, but that it was obviously nonhuman. It also stated that a U.S. signal Service officer believed that the pilot was from Mars. "The pilot's funeral will take place at noon tomorrow," said the article.

Many UFO "experts" have proclaimed this incident in Aurora to be a hoax, and with good reason. Aurora was a dying town which the railroads had passed by because a plague of yellow fever had killed off most of the people, and Aurora was quickly becoming a ghost town. Aurora residents who survived may well have concocted the incident to give themselves some publicity, and with the legitimate wave of "airship" sightings across the country, this seemed to be the ideal stunt.

However, in 1983, Walter Andrus, an investigator for the Mutual UFO Network (MUFON) in Texas reported that a piece of the mysterious airship has been found, and that tests have shown that it is not of terrestrial origin! Andrus claimed that a scrap of metal found near Proctor's welltower by MUFON investigator Bill Case conformed to the shape of the stone in which it was discovered, indicating that it was probably molten when it was hurled there.

Judge Procter's water well that was purportedly struck by an airship.

More importantly, X-rays verified the presence of pure aluminum in the sample. All commercial aluminum must contain some copper, but this sample was "pure". Andrus would not reveal the name of the lab that did the testing, however, because "they normally charge seventy-five dollars an hour". His group managed to avoid that fee.

One must still remain suspicious of this case, as I cannot imagine why no one has dug up the grave of the "Martian", which is marked by a "porthole" on the tombstone. Furthermore, no one can verify the lab tests on the molten metal. The Aurora "airship crash" is likely to remain a mystery, just like the airships seen cruising across the United States by thousands of witnesses.

This was not the end of the airship mystery, however, as more sightings continued after this date, including the Sistersville, West Virginia sighting above and one on April 21 in LeRoy, Kansas where a Mr. A. Hamilton reported a ship stealing some of his cows.

Were these sightings to come ten or fifteen years later, they might have described the airships as "zeppelins". However, the first rigid airship was not officially invented until 1900 by Count Ferdinand Graf von Zeppelin. The count ushered in the "Great Age of Airships" which lasted from roughly 1900 to 1937, when his zeppelin the Hindenburg burned at its mooring mast at Lakehurst, New Jersey.

The rash of sightings reported in 1897 and earlier are remarkably similiar to UFO reports today, fitting nicely into the "cigar-shaped UFO" catagory. Still, it is quite possible that many of the sightings were indeed of dirigibles of one sort or another, as the first power-driven balloon was invented by the French inventor Henri Giffard in 1852. The exploding airship in Aurora, Texas does not sound much like an "electro-magnetic-gravitational" powered ship, nor is it common for UFOs of the cigar shaped sort to go around dive-bombing cotton workers as in one of the earliest reports.

In the later part of the 1800's, balloons were becoming a popular form of transport, having been used by both sides during the American Civil War, and were the subject of a number of popular books, including Jules Verne's "Robur est Conquerant", H.G. Wells' "Around the World in 80 Days" and Maurice Renard's "Le Peril Bleu". Indeed, in the very year of 1897, a balloon expedition to the North

Drawing published by the British paper *Peterborough Citizen and Advertiser*, March 24, 1909. It shows the "aerial ship" observed by a constable, P.C. Kettle, "an absolutely trustworthy witness." Kettle's attention was drawn to the object by the noise it produced, "similar to that of a car." The object seemed to be equipped with a powerful searchlight.

Pole was being undertaken, under the leadership of August Andree. The expedition completely vanished and became a mystery until remains of the expedition were discovered on White Island in the Artic in 1930.

None of this can answer the question of the airships that flagrantly flew across the United States (and quite possibly other areas of world, England certainly) in the late 1800's. One possible answer is that a Captain Nemo-type inventor manufactured a rigid airship long before Zeppelin, perhaps one powered by steam-motors, or even electric motors of some sort. It is quite possible that some genius inventor even came up with some sort of "Anti-Gravity" drive for his private fleet.

Many of the descriptions fit perfectly into what are described as the type of ships that were first built in Atlantis and Rama some fifteen to twenty thousand years ago. Cigar-shaped airships used for cargo and troop carrying by Atlantis, purportedly, were called Vailixi by the Atlanteans and certain "occult" writers. The Indians of the Ancient Rama Empire, a civilization that existed parallel to Atlantis in Northern India, and of which a number of cities still exist today, used a similar airship known as a Vimana. Indian epics are full of detailed descriptions of these flying vehicles, of which there were many types, and even whole flight manuals are still extant, some even translated into English by publishers in India.

Eventually, Atlantis destroyed Rama in a horrendous war, carefully detailed in the ancient Indian epics of the Ramayana and Mahabarata. Shortly thereafter, Atlantis itself was destroyed in a geological upheaval that sank their mid-Atlantic continent, according to ancient Egyptian records. "Written history" begins just after this period.

Yet, what happened to the Vimanas and Vailixi of Rama and Atlantis? Reportedly, many were kept in use by secret "Brotherhoods" like the Essenes, Rosicrucians, Hermetics, etc. They were groups who kept old traditions, inventions and knowledge alive, much as the Masons claim to do today. Vailixi and Vimanas were reportly kept in secret fortresses in various places around the world, such as Tibet, the Alps and Mount Shasta in California. They get out their airships, driven undoubtedly by "electro-magnetic-gravitational" motors, and go places, visit their "brothers" in different parts of the world or make special trips to the moon or other places.

Assuming that this theory may be true, could some of the airship sightings that occurred "before there were airships" be sightings of such left-overs from Atlantis and Rama? The almost meandering "vacation-cruise" type of trip across the American mid-west in 1897 sounds almost like a bunch of holiday observers out to see how the great plains had changed over the past thirty years. I can almost envision a bunch of "Brothers", humans, mind-you, not aliens, picking up a Vimana at the "Mount Shasta Used Vimana and Vailix Lot" and going for a week's cruise. The airship exploding in Aurora, Texas, however, seems to be a horse of a different color.

Of course, as many UFO researchers would have us believe, these sightings could also just as easily have been alien ships, out for a similar purpose, though cylindrical-shaped airships are generally thought to be the large, "Interstellar" ships, and are not generally known to come too close to earth. It is also hard to imagine a bunch of irresponsible, joy-riding Martians dive-bombing frightened cotton workers.

Zeppelins became very popular in the early part of this century. Unfortunately, it was also found that they came in very handy for making war. The first large bombing raid ever staged on a civilian target was against Antwerp, Belgium by a single zeppelin in late 1914, during the first World War. Improvements in anti-aircraft weapons made such raids almost impossibly dangerous for hydrogen-filled airships by 1917.

The Germans, however attempted to stage one last bombing raid over London on October 19,

1917. Eleven, 700 foot long zeppelins with 2.3 million cubic feet of hydrogen each, flew over the English Channel at dusk and began a bombing cruise over the English Midlands, inflicting considerable damage. At midnight a storm blew them back across the Channel and they came under attack by Belgian anti-aircraft fire. Forced up to 20,000 feet in altitude, the crew nearly froze to death from cold and lack of oxygen. By dawn, six had made it back to Germany, two had exploded from the anti-aircraft fire and two made forced landings in enemy territory.

One ship was left adrift, and tried to fly over the Alps in an effort to get back to Germany. It crashed into a mountain peak and the control car and aft engine were torn from the body of the zeppelin. With the gas cells apparently still intact, the Captain and crew watched the body of the airship rise rapidly into the sky with as many as four crew members still on board. It completely vanished after this, which is still a mystery, as a 700 foot zeppelin is hard to hide, and considering its large aluminum frame, it could not have burned up entirely.

At the time of the Armistice in 1918, the huge new Zeppelin L-72 was nearly completed at the zeppelin works in Fredrikshafen. It had been planned that the zeppelins would have led an airship bombing raid on New York. However the war ended just before that. The L-72 was then handed over to the French as part of war repayments. The French renamed the ship the Dixmude.

The Dixmude set all kinds of new records and on December 18, 1923, it headed for North Africa to set new records with a crew of 50 under Commander du Plessis de Grenedah. On December 20 the Dixmude was spotted over Insalah, Algeria, deep in the Sahara desert. However, on December 29, the body of Commander du Plessis was fished off Sicily and his watch had stopped at 2:30. A stationmaster on the island claimed to have seen a light over the sea at 2:30 A.M. on December 23. On December 31, burned fragments of the control car were also found in the sea near Sicily.

This is an aerial photograph of a B-36 jet bomber. Because such bombers appear as thin black disks in photographs, they are often mistaken for UFOs.   U.S. AIR FORCE

It all made for a confusing story. How could the Dixmude have been spotted in Algeria on Dec. 23 and later spotted deep in the Sahara desert on Dec. 26? The wreckage of the zeppelin was never found either. Was it some other airship that was spotted in the Sahara? What was the cause of the Dixmude's demise anyway? One theory was that the ship broke into two (something that has been known to happen to zeppelins), and one half went down in Sicily, the other in Algeria. This hardly hardly seems satisfactory; it would seem that a phantom airship was cruising the Sahara desert or that there was some political espionage going on. Perhaps the Dixmude was deliberately sabotaged (much the same as the Hindenburg) and/or hijacked by some unknown group with some unknown purpose, perhaps with affinities to the French government itself.

The age of the giant, rigid airships was coming to an end. After the May, 1928 disaster of the airship Italia, which had flown over the North Pole and then crashed on an ice pack (the great Norwegian explorer Roald Amundsen vanished when he went to look for the survivors in a seaplane), other disasters befell American and British rigid airships and they were abandoned. Only the old, reliable Graf Zeppelins remained, and after the Hindenburg disaster, public confidence in zeppelins was utterly destroyed, as if it were a carefully mastered plan. The Nazi Government disbanded the Zeppelin works, and rigid airships were no more. All that remained (and still remain to this day) were the helium-inflated nonrigid airships (blimps) of the United States Navy.

On August 16, 1942, the U.S. Navy blimp L-8 left on an anti-submarine patrol out of Moffett Field near San Francisco, California. A small airship, it had a crew of two, including Ensign Hank Adams, a survivor of the crash of the Macon, a rigid airship that had burned up some year before. The blimp left its mast at 6:00 A.M. and reported seeing a oil slick at 7:30. No more radio contact was ever made with the blimp and search planes were sent out to search for it. At l0:30 A.M., an airliner sighted the blimp near San Francisco. Fifteen minutes later it came down on a beach near Fort Funston.

There was no doubt that something was amiss aboard
the Navy airship *L-8* as she returned from sea and
crossed the California coastline.
—U.S. Navy

Fishermen trying to catch the blimp noticed that the cabin door was open and that no one was on board. A gust of wind dragged the blimp along the beach, it struck a cliff and discharged a depth charge. Thus lightened, it rose again and drifted to the southeast. It came down on a street in Daly City, south of San Francisco, apparently because of a leak caused by the collision with the cliff.

The two crewmen were not on board, and all of the survival and rescue gear were still in the cabin. The regulation life jackets however were gone. No trace of either crewmember was ever found and the Navy could not give any reasonable explanation of the entire affair. It is conceivable that the men set the airship down on a beach (it was cloudy day with little visability) and abandoned the ship. Doing so without radioing would imply a sinister motive, and it is interesting that Ensign Hank Adams was the survivor of another "mysterious" (?) airship disaster.

After the *L-8* crash landed on a city street,
it was discovered that her crew had vanished. No clue to
their disappearance ever turned up.
—U.S. Navy

Zeppelins are extinct, and the blimp is certainly a vanishing breed, though Sunday football watchers will often see the Goodyear blimp hovering over a football field. Yet, "cylindrical, cigar-shaped airships", are reported all the time! Their resemblance to the vanished zeppelins of yesterday is uncanny to say the least, though the evidence is overwhelming that such airships, probably powered by "anti-gravity", existed long before the German Count persuaded some wealthy friends to invest in his ideas.

There is a nagging feeling though, that the demise of zeppelins was a planned, carefully carried out plot. But why? And by whom? Could some element of private business or a Military group in some government want to halt the scientific progress in "lighter-than-air-ships"? Yet, the rigid airship may not be dead. Several new designs are being experimented with, and cigar-shaped aircraft are ideal for using "electro-magnetic-gravitational motors".

Newspaper Headlines of the Past, Present & Future: Or "Aliens Stole My Baby"

WEEKLY WORLD NEWS 50¢

SAVED BY A GHOST! Little boy pulled from well by his dead mother

Crashed starship looks like solid gold, says diver

NAVY RAISES UFO FROM PUGET SOUND

Government cover-up after alien vessel is salvaged near Seattle

Starships vaporized as . . .

HUNDREDS WATCH UFO BATTLE

Hour-long dogfight blazes across sky and turns up on radar screens

★ EXTRA ★

THE DAILY TRIBUNE

FINAL
WALL STREET CLOSING
COMPLETE SPORTS
★ ★ ★
WEATHER: Continued Fair

Vol. 1, №7   "A Newspaper Dedicated to Worldwide Unity and Interest"   Latest News — Photos

GOVT. ADMITS ANTI-GRAVITY
"WE'VE BEEN MAKING FLYING SAUCERS FOR YEARS"
SAYS PRESIDENT IN NEWS CONFERENCE

Natural Foods Use For Products

STOCKTON, Calif. — The new low-calorie food industry is providing the nation's farmers an outlet for farm policy against that in new Tillie Lewis, founder and President of Tasti-Diet Products, Inc. Before the income of fruit and vegetable growers, Mrs. Lewis pointed out that the take of dietetic foods will top $100,000,000 this year compared to $25,000,000 only four years ago. Fruit and vegetables used as ingredients of these new updates represent a large percentage of farm products.

Panic In New York; Menagerie Breaks Loose

NEW YORK — Wild animals that escape from the circus shop or two often aren't as big as planting they take them into the human beings in the hopes of their chance meetings in the streets.

Take the case of the terrified tropical American wildcat which escaped from a Washington shop and roamed for many hours around strong houses. Chased by a squad of New York Chinatown updating police shotgun teams the wildcat was shot dead by his halted fire.

Much less trouble was Jackie a young male lion who escaped from New York Madison Square Garden. Jackie wandered into a basement and lay down for a short nap. He was easily caught in time for the circus.

A kangaroo down from a TV studio in Baltimore delayed the...

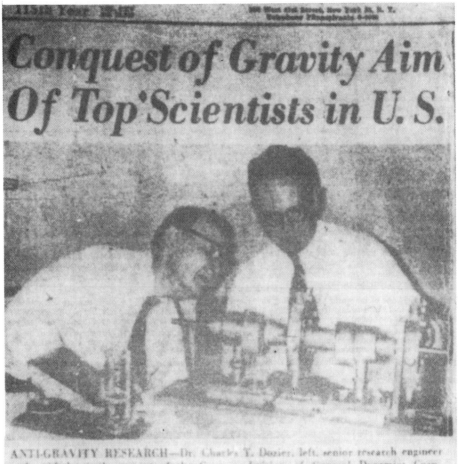

DUNLAP'S LAWS OF PHYSICS:

1. *Fact is solidified opinion.*
2. *Facts may weaken under extreme heat and pressure.*
3. *Truth is elastic.*

NEWS FLASH
PREHISTORIC BEAST ATTACKS!
City Ripped by Raging Sea-Giant From Ages Pas[t]

Mass Panic as UFO Terrorizes Villages

Thousands Watch in Horror As Strange Object Appears To Spread Mysterious Fires

Mysterious "cold" fires swept across dozens of villages in Russia, terrorizing thousands of people but causing no damage — and Soviet scientists say the phenomenon was caused by a giant UFO!

The bizarre flames engulfed a 3,000-square-mile area, causing "mass panic," reported the Soviet newspaper TRUD.

But as firemen raced to put out the fires, they discovered to their shock that "the flames kept retreating ahead of them," said Igor Vostrukhin, a special investigator for TRUD.

Four fire trucks chased the huge UFO — described as an orange-colored sphere 400 feet in diameter — but it, too, retreated ahead of them and finally shot straight up out of sight, said eyewitnesses.

SCARED OUT OF THEIR WITS: Estonian villagers flee their homes as firemen chase an eerie glowing object in this sketch depicting the bizarre UFO incident

It Was a Signal From Space, Says Top Astronomer

The incredible incidents occurred in the Soviet republic of Estonia on March 2, 1984. In a lengthy report, TRUD said. "Thousands of terror-stricken people spent an agonizing night watching flames shoot high into the dark sky across miles of Estonian flatlands... the night was filled with the mournful wailing of fire engines as fire trucks from over 20 major stations raced to the reported fires."

Vostrukhin added, "The flames kept eluding the firemen, remaining two to three miles ahead of them as though on the run from them.

"As a fire truck would enter a village that had been reported threatened by the flames, the firemen would find the place untouched by fire, its inhabitants scared out of their wits by flames shooting up and coloring the sky crimson. An invisible cloud of heavy odor hung in the air, causing local inhabitants to cough and making them nauseous — but they never felt or saw the fire over their own village that the firemen had responded to!"

Around 2 a.m. a fire truck approached a row of buildings that appeared to be on fire — when a UFO rose up from behind the buildings.

"According to the fire chief, the UFO was 400 feet in diameter and looked like a huge ball. Sparks and flames were dancing on its surface," said Prof. Agar Volke, a Soviet astronomer and head of the Estonian branch of the Soviet Commission on UFO Investigations.

"The next instant, the ball took off into the night."

The fire truck — soon joined by three others — followed the UFO, racing after it at speeds of up to 80 miles an hour. But they couldn't catch the UFO, said Prof. Volke.

"The firemen realized it was a spaceship. It was flying so low it was practically touching the treetops.

"The sphere moved over a swamp — and stopped. The firemen said they saw what they thought were round portholes, well over 100 of them."

Psychic Reseachers Say...
Your Grandparents May Be From Another Planet
...Here's How You Can Tell

You can tell if there is a space alien in your family tree, say psychic experts — who give tips on how to spot one.

"Many Americans have space alien ancestors, but they don't even realize it," says Brad Steiger, who with his wife Francie has studied UFOs, space aliens and outer space phenomena for over 20 years.

Think back to details you know about your grandparents and great-grandparents, suggested the Steigers. If one of your ancestors fits most, if not all, of the descriptions below, you probably have a space alien in your family past:

- **Unusual diet:** Ate strange foods, like mounds of cottage cheese, at unusual times, such as in the middle of the night.
- **Had dreams that came true:** Reported nightly dreams of events that later happened.
- **Unexplained history:** For example, went away for several months or longer without telling anybody where, and refused to talk about it afterward.
- **Generosity:** Gave money and time to charity; worked on neighborhood and school projects, etc.; cared about society, and was willing to work to improve life on Earth.
- **Sensitivity:** Had ESP, amazing psychic powers or was able to foretell the future with uncanny accuracy.
- **Ability to counsel others:** Everyone came to him or her for advice.
- **Led a charmed life:** Was extremely lucky; never got sick; won prizes in contests; frequently received gifts, and seemed to get money out of thin air.
- **Owned advanced equipment or hardware:** For example, was the first person around to buy a car or a TV set.
- **Acted childish at times:** Enjoyed playing children's games like marbles, for instance.
- **Appeared to be protected by unseen forces:** Escaped many dangerous situations that no one else could.
- **Talked of strange beings:** Said he conversed with spirits.
- **Had no past:** Talked of no family history or background.

— **LARRY HALEY**

Some people think the world is dirty without stopping to reflect that maybe they forgot to clean their eyeglasses.
— Billy Graham on The Dick Cavett Show

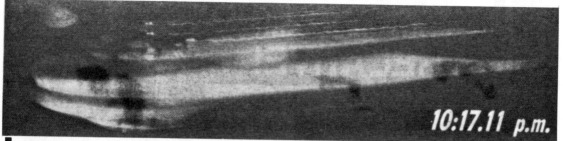

ARTIST'S CONCEPTION of the starships streaking across Kenya's nighttime sky.

Starship dogfight seen by hundreds

As hundreds of people watched in horror, three alien starships engaged in a spectacular dogfight and disintegrated in midair in what experts believe to be the first known battle between UFOs.

The stunning aerial assault was witnessed by more than two hundred African villagers, who watched for nearly an hour as the eerie scene unfolded, lighting up the night sky over Kenya.

French aviation expert Louis Bujon arrived in Kenya with a team of researchers one day after the incident occurred.

"Never before has there been a UFO encounter that was so well documented," Bujon said.

"We have confirmation from Cairo that the three crafts appeared on military and civilian radar screens and vanished simultaneously."

Bujon said his researchers have interviewed dozens of witnesses and will remain in the field until all those who saw the dramatic incident have been contacted.

"The eyewitness accounts that have been gathered to date are virtually identical in every aspect," said Bujon, a professor of astrophysics.

According to Bujon, villagers in the Rift Valley north of Mount Kenya noticed an unusual light in the sky at about 9:20 p.m.

"This white light descended rapidly toward earth and was at first thought to be a shooting star or some other naturally occurring phenomenon," the professor said.

"However, at a height of approximately 3,000 feet above the ground, the light stopped descending and hovered there for about 15 minutes. It was very still."

Bujon said that a second light suddenly appeared beside the first and both began a slow descent toward earth.

"By this time there were between 200 and 300 people watching from the ground," he said. "They described the approaching crafts as round, brightly lit objects. At this point, a third craft, much larger than the others, approached rapidly from the south."

According to eyewitness accounts, the larger UFO fired what appeared to be beams of light at the other two crafts, which veered away in opposite directions.

"The two smaller saucers circled around and returned fire with what we can only assume to be some type of laser weaponry from the descriptions that we have received," Bujon commented.

"At precisely 10:17 p.m. the three crafts disintegrated in a tremendous explosion."

Bujon said he does not believe the explanation offered by some scientists that what the villagers actually witnessed was a meteor shower.

"Evidence does not support that conclusion," he said.

"This was a skirmish in an interplanetary war."

— MARC JUBERT

Hour-long battle blazes across the sky and turns up on radar screens

SPECTACULAR explosions ended an amazing aerial battle, according to eyewitnesses.

Long-winded woodwind

Joe Silmon of Gosport, England, played the flute for 48 hours in 1977. According to the *Guinness Book of World Records*, that's the longest recorded marathon by a flautist.

WEEKLY WORLD NEWS
July 30, 1985

1 in 200 Americans is from outer space

...claims UFO expert

ONE OUT OF every 200 Americans comes from outer space! According to experts, these aliens arrived from a number of different planets and solar systems. They have chosen the Earth for a home because of its atmosphere and warmth.

The United States has become the favorite stomping grounds for these space aliens because it is so easy for them to travel and work here. This is one country where citizens are not required to show government-issued identification papers wherever they go.

According the Edward Huhn, director of an extraterrestrial study group, these beings are examining our planet and studying us as individuals.

Huhn says he first began researching the presence of space aliens in the United States in 1980 when he found a crashed UFO in the countryside outside of Ketchum, ID.

"Inside the spaceship were the remains of three aliens," reports Huhn. "There were also outfits of synthetic skin, and we found three sets of false identification locked inside of a thick, plastic box."

Huhn says that was when he formed a team of UFO investigators to find out how many of these individuals might actually be infiltrating the U.S.

"Most of these people have no past, no record of their birth, no family members, no nothing," says Huhn.

"They just move into town and begin to exist."

The researcher says whenever he or one of his fellow investigators makes contact with one of these space aliens, the alien disappears the next day.

"One man we contacted in Cranford, NJ., disappeared a few hours after we met with him," says Ketchum. "We were granted access to his abandoned apartment. There we found a series of notebooks which contained observations of the daily routines of human life. It was spooky."

According to Ketchum, as many as 5 million individuals living in the United States are actually visitors from another planet. Many of these visitors are married and have children.

The researcher notes that when a space alien feels comfortable and undetected on this planet, it appears to have no desire to return to its native planet.

August 6, 1985 — SUN

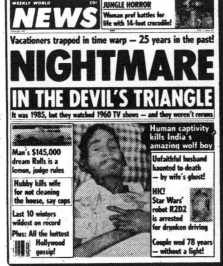

Sun

Vol. 3 — No. 35 August 27, 1985 55¢

The terrible curse that hangs over Bill Cosby's head

Mother of child with pointed ears tells how...
I GAVE BIRTH TO UFO BABY

UFOs Are Blasting Our Planes Out of the Sky

...Says Expert

UFOs are real — and their actions have become increasingly violent, says a leading UFO expert.

"Flying saucers aren't science fiction. Dozens of military aircraft have been shot down or literally blown away by UFOs," declared retired Maj. Hans Petersen of the Danish Air Force.

"And now there is clear evidence that these visitors are showing a more aggressive attitude toward Earth."

Maj. Petersen, an air traffic expert, has collected a 5,000-page file covering more than 50 years of UFO activity. His material is based on classified military reports, secret government documents and personal contacts with officials of NATO and various air forces.

One astounding case from his files involves a Russian jet fighter that encountered a UFO over Cuba in 1967.

"The UFO was described by the pilot as a 'bright

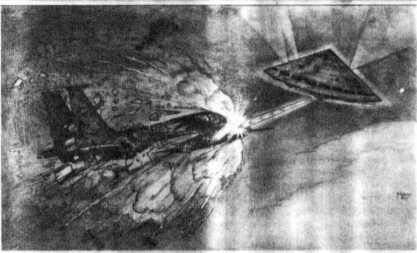

LASER BEAM from UFO shaped like a Chinese coolie's hat gunned down a U.S. F-15 fighter jet.

metallic object, like a spherical envelope,'" said Maj. Petersen, 61.

"The pilot was given the order to attack and destroy — but seconds later a scream was heard over the radio by the air traffic controller, and the jet exploded."

Another case from Maj. Petersen's files dates from June 9, 1974, when a jet from the Japanese Defense Forces was ordered to attack a UFO that had violated Japanese airspace near Tokyo.

"The jet was suddenly hit by a laser-type beam, and it disintegrated.

"In another incident, on Aug. 26, 1984, a squadron of Russian fighters were scrambled to intercept a UFO over Lithuania. The astonishing thing is that they were ordered to attack on sight.

"The squadron leader announced over the radio that he'd spotted the UFO and alerted his squadron to pre-

'Flying Saucers Aren't Science Fiction — We've Shot at Them and Now They're Shooting Back'

Hans Petersen

pare an assault. But the next second there was a blinding flash of light and the lead jet simply evaporated. The attack was immediately called off. And the pilot's body was never recovered."

Another recent case from Maj. Petersen's files concerns a U.S. F-15 fighter that encountered a UFO near Alaska.

"The U.S. pilot reported visual contact with a UFO that was 'shaped like a Chinese coolie's hat,' with flashing lights.

"The pilot's transmission to his base was interrupted by a strange but humanlike voice, warning the pilot to leave the airspace within 10 seconds. Instead, on orders from his base, the pilot attacked the UFO, firing eight missiles.

"The missiles all exploded harmlessly against some kind of force field, half a mile from the UFO, and the UFO countered with a laser beam that blasted the U.S. jet out of the sky.

"In all the cases from my files, the incidents have been cleverly covered up by the authorities," added Maj. Petersen.

"But the material clearly proves that UFOs are space ships from other worlds. We've been shooting at them — and now they're shooting back!"

Lieut. Gen. Bent Amled, chief of staff at the NATO Northern Command Base in Norway, said: "Maj. Petersen is a respected ex-officer and a serious UFO researcher."

— WILLIAM DICK

PENTAGON COVER-UP?

UFO crash mystery

A submerged UFO, found off the U.S. coast by two divers, has mysteriously vanished. But there is growing suspicion that the downed alien craft was recovered — in a secret cover-up mission carried out by the Navy!

The metallic, dome-shaped craft was found in the exact spot where witnesses earlier reported seeing a "glowing" UFO crash into the frigid waters of Puget Sound off Bellingham, Wash.

The witnesses said the spacecraft hurled a shower of sparks and flame at least 70 feet into the air and turned the water into a boiling froth of foam.

A few months later, Seattle divers John Walker and Richard Burke made three dives in the exact spot where the UFO slammed into the water, and both reported finding a large, dome-shaped object, somewhat golden in color, lying half buried in the mud 180 feet below the surface.

Despite its eerie appearance in the turbulent, murky depths, the sunken craft brought shouts of excitement from Burke.

"Whooooeeee! I'm standing on the UFO," he yelled into his helmet intercom. "It's pretty damn big."

Burke described the object as being about 17 feet in diameter and looking like an inverted teacup half-buried in the mud at about a 45-degree angle.

"When I stood on it I could hear a low-pitched humming sound," he said. "It made us very nervous. It seemed to be about 10 feet high, but we couldn't really tell because most of it was buried in the mud."

Burke and Walker made two dives on two consecutive days. After a three-day interval, they went back to recover the UFO. But this time, they were unable to find it.

"Today, we have equipment that can pinpoint a place on the ocean floor," said research scientist Dale Goudie, who had taken part in the effort to salvage the UFO. "We knew exactly where the object was. It had been there for five months without moving. Why should it suddenly disappear? We have heard reports that the Navy went out and picked it up."

Amid growing suspicion of a Pentagon cover-up, government officials have denied all knowledge of the UFO and have refused to comment.

"I don't know if we have the object or not," declared a high-ranking Navy official in Washington, D.C. "If we do, it may be classified. It could be a military device."

Defense Department physicist Bruce Maccavee — who also happens to be the chief of the Fund for UFO Research — said the dome-shaped object could be "highly significant UFO evidence" which may never be revealed to the public.

"Puget Sound is loaded with naval facilities," Dr. Maccavee told The NEWS. "I would think if the Navy wanted to recover the object, it could have been done without anyone knowing about it.

"My gut feeling is that there was something there and it just disappeared. If it was just a piece of junk, it should still be there. But if the Navy does have it, I don't know how you can get them to admit that they do."

Witnesses say impact hurled sparks & flame 70 feet in air

Did a Navy team secretly recover wreckage found by two divers?

APRIL 30, 1985 - SUN - PG. 17

UFOs attack Moscow — 100s kidnapped

THE AIR RAID sirens in Moscow are no civil defense exercise!

The clear and present danger is from hostile UFOs which are kidnapping hundreds of people, making them vanish without a trace into the silent sky.

Every twinkling star is viewed with suspicion, for no one knows when one will turn out to be a lurking UFO, which will suddenly descend to grab another trembling victim.

Beams

Terrified witnesses report watching helplessly as humming UFOs emit beams of light that lift people off the ground and bring them up into giant ships, never to be seen again.

Soviet authorities are warning people to stay indoors whenever possible until the threat is past, but no one knows if the menace ever will end.

While people talk about little else, the Soviet media has downplayed the seriousness of the situation, mentioning the sightings only in an occasional back-page article of their newspapers.

The sinister snatchings began recently, as a UFO ripped into Soviet airspace from the Artic Circle at more than 600 miles per hour.

As the UFO sped from Gorky to Moscow, and refused to answer requests to identify itself, MiGs were scrambled to intercept and destroy the mysterious intruder.

As the many blips of the Soviet planes approached the airborne stranger, the UFO image on radar split into four separate dots, confusing the trackers.

Swedish radar technicians watched as all six ultra-modern Soviet craft were lost in a one-sided, three-minute clash with invaders from outer space.

Vanquished

The Soviet state is powerless, its best aircraft vanquished. The UFOs slip in at will and pluck people whenever they please.

Witnesses in Gorky, who saw the alien object from the ground, were reported in Trud, the trade union daily paper, as saying, "It was long, grey, shiny, shaped like a cigar, with no wings or tailfin."

The UFO was seen by thousands, "striking terror in the hearts of those who saw it," Trud informs.

Vladimir R., a Moscow steel worker, told Western journalists, "A spinning disc about 100 feet wide hovered above the plant where I work. It gave off a humming noise.

"A beam of light came out from its bottom and caught two of my fellow laborers. They were raised into the air and brought into the spaceship, which then zoomed out of sight."

Soviet authorities do not know how many people have already been kidnapped, but the total could well reach into the hundreds."

— PAUL HOWARD

N.Y. Police Watch Huge UFO Buzz Nuclear Power Plant

By FRANKLIN R. RUEHL

An American nuclear power plant was buzzed by an enormous UFO, say the plant's security guards, local police officers and other eyewitnesses — and, incredibly, the New York Power Authority confirms the sighting.

"On the night of July 24, 1984, a UFO was seen by security guards over the Indian Point nuclear power plant," said Carl Patrick, spokesman for the New York Power Authority, which operates the plant near Peekskill.

"But the UFO in no way affected plant operations."

UFO investigator Philip Imbrogno noted:

"This was the first time ever that there is confirmation that a UFO was sighted over a nuclear power plant."

Imbrogno, who's associated with Dr. J. Allen Hynek's Center for UFO Studies in Arizona, interviewed six of 12 guards who saw the craft.

"They said the UFO was gargantuan and looked diamond-shaped, measuring about 450 feet in length. At first, it was bright white. But at one point it changed from white to blue to red to green to amber," Imbrogno said.

"It hovered above the plant for approximately 15 minutes, from 10:30 to 10:45 p.m. It was roughly 30 stories above the exhaust funnel of one of the plant's three nuclear reactors."

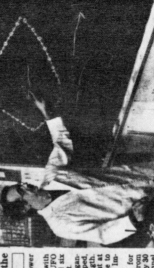

INVESTIGATOR Philip Imbrogno sketches drawing of UFO on a blackboard.

Many other people in the area also reported sighting the UFO.

"Local police in Peekskill received numerous calls reporting UFOs during that evening.

"Many officers from various police stations personally told me that they saw the UFO," Imbrogno added.

Sgt. Karl Hoffman of the Peekskill Police Department admitted: "On the evening of July 24, 1984, I observed an unidentified flying object. I saw a dozen white lights in V-formation, slowly moving toward the power plant at Indian Point.

"I have absolutely no idea what that object was."

Television is like a toaster. You push a button and the same thing pops up every time.
— Alfred Hitchcock

POWER PLANT wasn't affected by UFO visit.

UFOS are from Earth & crewed by humans

UFOs come from Earth, not other planets — and they're manned by people, not extraterrestrials, says a leading expert on aliens and their spacecraft.

"I'm convinced they're making their craft on this planet — and have been doing so for years," says Laurie Campbell, who has studied UFOs for 20 years.

He believes that governments around the world are secretly working with the aliens — and are trying to suppress this information.

"It is awkward to talk about it as it smacks of world government. Every time I lecture about this, or publicize it, my phone is bugged or I come under surveillance," he says.

Campbell, the acting president of the Perth UFO Research Group in Australia, says he has had contact with the aliens both in Australia and overseas three times.

In fact, aliens have told him telepathically how to make a spacecraft of his own — one that Campbell hopes will take him to Mars in 15 years time.

"I've been accused of imagining or hallucinating, but when contact is made you know it's pretty real."

Campbell has put the information gleaned from the aliens in these encounters to good use by making several space models.

His latest model is the Cygnus II, which he says could be in service in 15 years.

LAURIE CAMPBELL is making a spacecraft that will take him to Mars.

By GUS VANDERMEER

Before his meeting with the aliens, Campbell admits he was hopeless with his hands. But now he has ideas and expertise to construrct the models, "as if someone was putting them in my head."

The dauntless researcher has been investigating UFOs since 1955. He now wants to set up a traveling space show to bring the truth of UFOs to the general public.

August 27, '85/EXAMINER

Levitation:
Personal Defiance of Gravity

by David Hatcher Childress

While the concept of Anti-Gravity may be relativly new, people have been personally defying gravity for thousands of years. Incidents of levitation have been recorded in many ancient Hindu, Babylonian, Chinese and Egyptian texts, but they may be considered a bit distant and second hand for serious consideration.

Probably the earliest and most famous of levitations worth considering is that of Jesus and his disciple Peter walking across the water on the Sea of Galilee. While many people have demonstrated remarkable buoyancy in water, it is quite uncommon, to say the least, to see someone actually walking on water. Peter, we can suppose was able to walk on water largely because of his faith in his benefactor and himself, as Jesus suggested that he walk across the water and join him. Jesus then later levitated before his disciples after his cruxifiction, ascending, as it were, to heaven.

Other incidents of levitation, starting in the middle ages, are fairly numerous. Saint Francis of Assisi, sometime after he received the stigmata on Mount Alverna in 1224, began occasionally levitating. A friend of Saint Francis', a Brother Leo, often visited the Saint on Mount Alverna and testified to seeing the Saint levitate on numerous occasions, once 3 cubits in the air, 4 cubits on another time, on another to the height of a beech tree and on one occasion, Saint Francis was lifted up so high and surrounded with such splendour, that Brother Leo scarce could see him!

Saint Teresa of Avila was a Carmelite nun who died in 1582 and was given to rapture and levitation frequently. On several occasions she levitated during full view of the nuns and bishops during communion, much to her own horror. In her autobiography, "Life", published in 1565, she describes her "rapture": "It seemed to me, when I tried to make some resistance, as if a great force beneath my feet lifted me up...I have to say that , when the rapture was over, my body seemed frequently to be buoyant, as if all the weight had departed from it; so much so that now and then I scarcely knew that my feet touched the ground."

Francisco Suarez (1548-1617) was a Spanish member of the Society of Jesus and one of the great theologians of the Roman Catholic Church, and apparently a frequent levitator. He did his levitation in the secret confines of his monastic cell, but was discovered one day by a Brother Jerome who filed an affidavit stating that he witnessed Francisco hovering in the air about three feet off the ground while in a kneeling position, praying to a cruxifix. Francisco made Brother Jerome swear not to tell anyone what he had seen until his death.

Maria Coronel of Agreda (1602-1665) was another nun who frequently levitated while in a trance state. Her fellow nuns occasionally exibited her to vistors while in her "levitation trance". She typically remained hovering a foot or so above the ground, often for several hours. She often resisted her ecstatic levitations so fiercely that she sometimes vomited blood.

Of all the levitating Christian Saints, probably the most unusual and best known, was Joseph Desa, canonized as Saint Joseph of Cupertino. He was born of poor parents in southern Italy in 1603 and took to an unusually harsh practice of austerity at an early age. He was in and out of several monastic orders in his youth, being dismissed for his "absentmindedness, fits of ecstasy and careless ways".

However, he eventually became an ordained priest on March 28,1628. His fits of ecstasy continued and in one instance, after Joseph had said Mass in a private chapel and had then withdrawn to a corner of the church to pray, suddenly, and without warning, he rose into the air and uttering a sharp cry, flew to the altar, his body upright and his arms outstretched. Seeing him alight on the altar amid the burning candles, several nuns began to scream: "He will catch on fire!" But Joseph's companion, Brother Lodovico, who seemed to have had some familiarity with such sights urged the nuns to greater faith and assured them that Joseph would not be burned. And sure enough, after a short time Joseph gave another cry and flew back from the altar, this time in a kneeling position, in which he landed safely on the church floor.

After a period of self doubt, when he was called to Rome and admonished for his behavior, he began a period when his ecstasies and levitations became his normal behavior. Music, in particular was apt to provoke an ecstatic flight. On one Christmas Eve, Joseph was so moved by the organ music in the church, he flew some 20 yards to the high altar, where he embraced the tabernacle and knelt for 15 minutes or so amid the burning candles. Then he flew back to the floor.

Saint Joseph's accounts of ecstatic levitations are far too numerous to go into in this book, but some of his other levitations are rather amusing. Once he flew up to the top of an olive tree after giving an ecstatic cry and remained kneeling on a thin branch for a half an hour or so. His rapture deserted him before he regained the ground, however, and a priest had to get a ladder to help him down. On another occasion, when 10 workmen had difficulty erecting a heavy cross near his hometown, he "rose like a bird into the air," easily lifted the cross, and set it in place! Once, Joseph cured a madman by grabbing him by the hair and levitating with him for 15 minutes or so.

St. Joseph of Cupertino taking off.

Joseph died on September 18, 1663 and just over a 100 years later, he was canonized a Saint.

Closer to our own time, the "spiritual medium" Daniel Dunglas Home was observed to levitate numerous times over a period of 40 years and was never discovered in any fraud. An article in the Hartford Times in the year 1852 stated that he levitated several times and was carried to the ceiling in a spontaneous bit of levitation at the age of 19. Later, he learned to control his levitations and did them for such notables as the emperor Napoleon III and Mark Twain. On one occasion in

Daniel Dunglas Home

Daniel Dunglas Home, among the most sought-after mediums of his day, was well known for his levitations before groups of witnesses. An eminent scientist who tested Home's extraordinary performance found no fraud.

1868 while at Lord Adare's home in London, England, Daniel, while in a trance, flew out of the window and came back in another window, doing this several times. Daniel Dunglas Home's popularity became enormous, and he moved in very aristocratic circles, particularly in Britain.

The Indian rope trick is one of the most well known magic tricks in the world, usually perpetrated by throwing the rope into the air so that it catches by means of a small hook, an invisible cord strung between two trees, so that the fakir may then climb the rope. An unusual account of an Indian fakir's levitation was published in the Illustrated London News in 1936 by P. T. Plunkett, a tea planter who observed a levitation in southern India.

The fakir, Slubbayah Pullavar, had "long hair, a drooping moustache and a wild look in his eye." After sprinkling water around a tent, he entered it and after a few minutes when the tent was removed the yogi was observed lying horizontally in the air, his hand resting lightly on top of a cloth-wrapped stick. Plunkett and his friends photographed the event and then walked around the yogi, passing their hands beneath him. Apart from the stick "the man had no support whatsoever."

About four minutes later the tent was put up around the yogi again, but the fabric was thin and Plunkett saw Pullavar's descent: "After about a minute he appeared to sway and then very slowly began to descend, still in a horizontal position. He took about five minutes to move from the top of the stick to the ground, a distance of about three feet...When Slubbayah was back on the ground his assistants carried him over to ...(us) and asked if we would try to bend his limbs. Even with assistance we were unable to do so." The yogi was doused with cold water and came to. It is interesting to note the immovably stiff limbs, as this is what may have held the yogi in the air, rather than genuine levitation.

Today, levitation still makes the news. The followers of Maharishi Maharesh have theoretically been instructed with a special "mantra" that allows them to levitate and even fly! Indeed, the Transendental Meditation folks have even released some photos of some of their followers 'levitating" while sitting cross legged in meditation. The Maharishi has even established his Maharishi International University in Fairfield, Iowa, where two meditation domes have been constructed so that meditators can safely, and privately meditate and levitate inside them.

A German film crew filmed an African witch-doctor in Togo in 1984 for the film, "Journey Into the Beyond". The three man film crew was led to a remote village in Togo where a sorcerer lived whom natives said was able to "fly". The sorcerer, Nana Owaka, was convinced to demonstrate his feat and then meditated for a full day. He then placed dried leaves and twigs in a circle and sat in the middle. "Just as the sun was setting, Owaka started to stir. A villager lit the circle of twigs and flames shot up. Drums began beating wildly- then we were hardly able to believe our eyes as Owaka stood and rose straight upward!.

"It was if he were being lifted on a pillow of air. He simply hung as if suspended, with nothing above or below him." The camera crew filmed him from two angles and "the levitation lasted almost one minute. Then, just as unexpectedly as it had begun, Owaka fell to the earth. According to Father Andreas Resch, a Catholic priest and director of the Institute of Frontiers of Sciences in Innsbruck, Austria, "I have invited different teachers of physics to look at the film and not one has been able to detect anything artificial which would have made this sorcerer levitate. I think the best proof for levitation to date is this film!"

'We Were Hardly Able to Believe Our Eyes as He Stood and Rose Straight Upward!'

Anti-Gravity Comix and Classified Ads

The Australian Spaceport—Ron Turner's dramatic depiction of the galaxy's largest space drome. (Frame from Super Detective Library #133 (1958), p.38. ©IPC Magazines Limited 1958.)

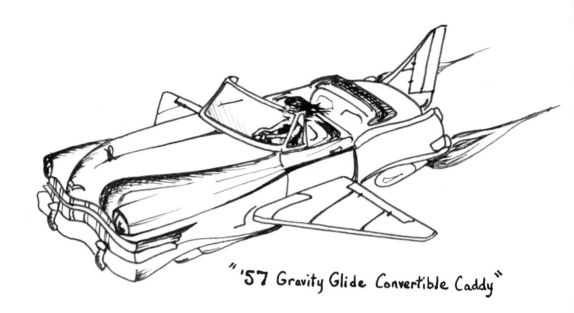

"'57 Gravity Glide Convertible Caddy"

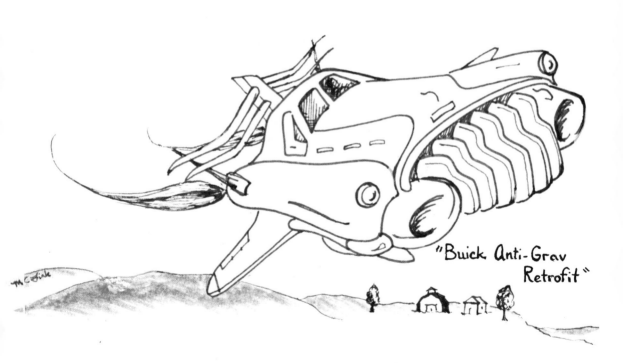

"Buick Anti-Grav Retrofit"

Tiger
by Bud Blake

SPACE, ASTRONAUTS

BOOMERANGS ... WORLD'S BEST: Incredible returns, scientific sport. **FREE** Illustrated catalog ... 70 boomerangs, books, plans. Write today! BoomerangMan, 1806-S N. 3rd., Monroe, Louisiana 71201-4222

LEVITATION by non mechanical means. Detailed information, diagrams. Send $20 cash, check, money order. 8-12 weeks. SpectraCom, PO Box 37-A2, Taylor, ND 58656

OFFICIAL NASA Cap. Send $9.95 to: "SHUTTLE" PO BOX 5108, Titusville, FL 32783

APOLLO, Shuttle, Voyager, NASA mission videocassettes. Free Catalog. SPECTRUM Video, Dept. SD, PO Box 5898 Unit 13, CA 91761

30 ANTI-GRAVITY Methods. Database Catalog, $1.00, Guaranteed. Rexfax, Box 1258, Berkeley, CA 94704

SPACENEWS CAPSULES. Shuttle, experiments, commercialization. 12 issues, $14.00. Sample $1.00. QW Communications, PO Box 1272, Waltham, MA 02254

ASTRONAUTICS

ANTIGRAVITY Propulsion Device. Free Brochure, RDA, Box 873-NS, Concord, NC 28025.

SPACE PEN—exact model of pen used by Astronauts and $10.00 to: H-Space Box 3053, Torrance,

ANTIGRAVITY AND UFO'S PROPULSION SYSTEMS RESEARCH COMPENDIUM. Free Brochure. HEER, P.O. BOX 5286, Springfield, VA 22150. Introductory Volume $20.00 (190 pages). Volume I $65.00 (450 pages).

Catalogue of **SPACE ART** and **SHUTTLE PHOTOGRAPHS.** 726-C North Milwaukee Street, Milwaukee, WI 53202.

SCIENCE AND CHEMISTRY

225 PAGES! 1984! GLASSWARE—CHEMICALS—LABORATORIES/BIOLOGY SUPPLIES—ESTABLISHED 1970! $3.00 REFUNDABLE! ALDRICH SCIENTIFIC, BOX 675-62SD, HELOTES, TX 78023

LABORATORY CHEMICALS, GLASSWARE—APPARATUS Catalog $1.00 HAGENOW LABORATORIES, 1302 Washington, MANITOWOC, WISCONSIN 54220

ELECTRO-GRAVITATIONAL Propulsion Energy Technology; Secret Documents, $10, 4735 Ravenna, Seattle, 98105

INDUSTRIAL CHEMICAL & SUPPLY CO., INC. REAGENT

OFFICIAL NASA Cap. Send $9.95 to: "SHUTTLE" PO BOX 5108, Titusville, FL 32783

SPACESHIPS and **SPACE DRIVES.** Free Brochure. Space Box 228, Kingston Springs, TN 37082

SCIENCE AND CHEMISTRY

BUILD YOUR OWN ROCKET MOTORS!
* 40 POUNDS THRUST!
* FIFTY CENTS EACH!

DANNY DUNN
and the
ANTI-GRAVITY PAINT

By Jay Williams & Raymond Abrashkin

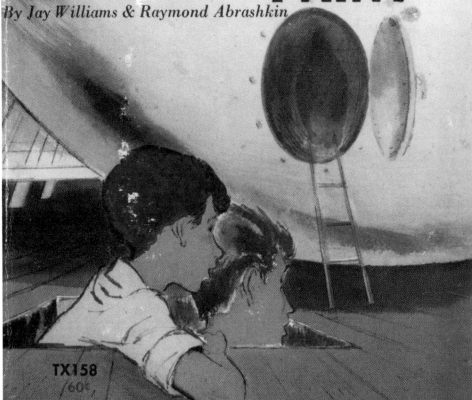

TX158
60¢

DANNY DUNN and the ANTI-GRAVITY PAINT

He reached up and caught hold of the Professor's ankles.

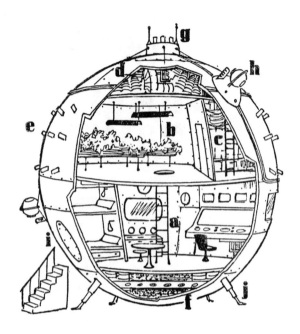

Diagram of First Anti-Gravity Space Ship

a Main living quarters—galley, bunks, control panel, ports, etc.
b Hydroponic garden
c Storeroom
d Machinery for heating, air circulation, etc.
e Belt of rocket exhausts
f Weighted base containing fuel tanks pumps, interior electrical system
g Solar battery & antennae
h TV system on mobile tracks
i Hatch entry
j Telescopic legs

Copyright © 1956 by Jay Williams and Raymond Abrashkin.

Meanwhile— Back In the Future...

by W. P. Donavan

As long as I'm off on the plane of wild speculation there's another point I'd like to look at, that to my knowledge has not been examined yet. What would happen if and when all this stuff eventually gets used in a civilization? After all, this society is scarcely above slightly organized barbarism. So let's reverse the polarity on the wayback machine and take a peek into the future. Let's set the year for say, 2491, and see the world through the eyes of a citizen in that time.

Ah, it's another beautiful day. I rise from a restful three hour sleep, thanks to the psychotronic system built into the bed. My bed may seem old fashioned but I'm into antiques anyway. The new ones look like a half-cylinder and have a gravity cancelling system built right into it so you sleep in free-fall. It's supposed to be a very relaxing sleep. Morning meditation doesn't open any insights. I throw a robe on and walk over to the kitchen and punch up breakfast. There's the usual soft buzz and drone as it reconstructs the food from the energy patterns stored in memory. I do some of my own cooking from time to time and save whatever tastes and looks good in memory. And now for a taste... Yech! (cough, cough). Oh well. Nobody's perfect.

You wouldn't recognize this cooktop. It looks like a white ceramic rectangle measuring 50 by 100 centimeters. The cooktop generates an energy bottle above it and has a programmable gravity field. So you can either cook in free-fall (which makes some of the wildest results you've ever seen), or crank the field density up to 100 times the earth's gravity field. I visited a museum once and took a look at something called a refrigerator. It's hard to believe that people allowed something that cumbersome in their homes. The thing looked intimidating. The Nutrimatic food materializer uses a pretty simple process, after all. It just generates a triplex wave hologram and pours raw energy into it from the system in the house. That system, by the way, is a solid-state free energy generation system. You might call it a spatio-temporal reactor. It changes the structure of a small volume of free space and literally wrings the energy out of it like a sponge. No pollution or radiation (Punch in a little cinnamon...here.)

The same triplex wave hologram is used in the life insurance policy that we subscribe to. Subscribers usually go in for an update about once a month. That way, if you have a potentially lethal accident (which is very rare now) they pop you in the regeneration chamber and you're as good as new. Now they're talking about a regeneration system which includes a genetic randomizer. That means that every thirty years or so you could take up the option and get a new body. That's causing a major uproar in this time. Life insurance seems like it's rapidly becoming outmoded as the lifespan streches out --somewhat disconcertingly, as the earth's population is rapidly expanding.

Each room in the house has its' own gravity control. I like to exercise in 2g. A sauna or steambath in zero-gravity is very relaxing. There's usually a fan to keep the air circulating so you wouldn't suffocate if you dozed off. My bedroom is usually set to whatever I prefer. One day I'll sleep in earth-normal. If I'm really bushed, as I was last night, I'll turn it down to a tenth of a g. Of course accidents do happen. One day the control shorted out in the rec. room and plastered the guests into the couch at 3 gees. I wondered what was happening when I looked into the room from the hall and saw the look of horror upon all their faces as their bodies were stuck to the couch with an iron grip. Of course I have a failsafe circuit in it now.

Moving furniture is a snap. Just turn down the gravity in that particular room. People who aren't used to that method and try it for the first time have quite a few accidents. Usually bruised ankles and smashed kneecaps. What they never count on is the inertia of the pieces that they move around. You sometimes see pockmarks on the ceiling after they're finished.

The carport would look familiar, except for a few things. My car is completely solid-state. The old models used bulky outboard turbogenerators but this one runs on a miniature spatio-temporal reactor. Very compact as far as a power supply goes (about 10 centimeters on a side). Lets see...how do I describe the way it looks? Imagine a corvette without wheels and completely streamlined for a top speed of 300 kilometers per hour.

I think that the major improvement was when they got rid of the wheels. Instead there are four electrogravitic impellers that measure about 30 by 30 cm. All four are used for levitation and keep the car above the ground when the car is at a full stop. Actually all they do is generate a negative gravity field that pushes the ground away and lifts the car up (thanks to Sir Isaac Newton). Each are shaped like upside-down, truncated pyramids. The pyramid faces are at 45 degree angles in relation to the ground and are completely self-contained within the body of the car. To accellerate, the sides facing the rear of the car are biased. Braking biases the front faces. To make a right-hand turn the right two impellers have a slightly lower overall field strength than the left two. The left hand faces of the two in the front are biased as well as the right hand faces on the two in the rear half of the car. There are no bumpers on the car. Who needs them when you have an integral deflection shield? Think of them as electrogravitic bumpers. The field strength is proportional to the speed of the car. To test them you tap the test button and when the indicator light comes on, hop out of the car. Then with the car at a full stop take your hands (palms out) and push against the field. The field should feel like soft rubber at a meters' distance and hard rubber at half a meter. The new model has inertia-absorbing brakes, which is an additional safety feature. The teen-agers use that to go roaring like a bat out of hell and make right angle turns at 300 k.p.h.. They think it's neat. I think they'd better watch it. My model has gravity restraints built right into the seats. In a collision it holds your body in the seat with a field that varies with the deceleration of the vehicle. It sure beats air bags!

Of course my family have a transport subscription. This is for the people who don't want to deal with the hassles of the skyways. Each house (or most of them anyway) have a transporter. In the past one company owned all the equipment and you leased it from them. Recently the company relinquished the ownership of the transporter units and now just deals with long distance runs. Everybody was in a very confused state just after the breakup. The company is called Atlantic Teletransport and Telecommunications. The heart of the transporter, the transtator, is just a superregenerative gravitational amplifier with positive hyperspatial feedback.

I'll try not to go into too much detail on the system since the only ones who would be interested in them are the ones who work on them. Basically it sets up a scalar black hole with the business end extending to wherever the coordinates are programmed to. You see, the only reason that a black hole (we call them CVE's, or charged vacuum envoidments) can suck in matter is because they have a gradient to their field. The transtator produces a field without a gradient to it. So the field just sits there and distorts space, warps it. The transtator is also a critical component in warp drive systems in the starships. You set the coordinates to the point to where you want to go from a transport book. The sweep field then goes through hyperspace and checks the status of the destination point to prevent an apport. Then you step onto the transmat and five seconds later you're there. We can go wherever we want to go for the duration of our subscription. The transport card that we have must be used for each transport, to keep track of where and when we go. This is because of a few unethical individuals in the early days who used it to beam into safes and remove whatever they wanted. Recently a few technicians who tinkered with the systems found out how to put a hyperspatial skew to the field which would give you a variable time vector, so you could beam into a point into the future or past and return at a prearranged time. This gives a lot of people the willies. It's such a recent development that I don't think anyone has tried it yet.

Our economy has gone through a major trauma since the utilization of free energy systems. You see, since energy was free and the technology to duplicate any commodity was readily available, an economy based on a commodity was impossible. In other words, if you had cheap energy and the means to convert that energy to mass--for example gold--then you have cheap gold. The same would exist for any other commodity. So we evolved an economy based on an energy which *is* limited; human energy! The system is based on caloric expenditure of the human system per unit of time. Any stress on the system would increase the 'burn' above the base level which is computed on an individual level. The unit is called the credit, which is kilogram-calories burned per task. It's the only system which is equitable and fair--at least at this time.

        I guess our values have changed quite a bit as well. The reason that society had materialistic ideals in the past was that material commodities were in short supply, and thus were valuable. In this case matter and energy is cheap--but thought is not. Now the only thing that remains valuable is knowledge, and that means that the individuals who have true wisdom are rich. That is recognized as such. This is not a utopia however, at least by my standards, and each of us struggles in our own way toward our individual ideals. Relationships are no longer on an exclusively materialistic level or purely spiritual, but on a mix of the two, more of a balance. Matter is recognized as form rather than substance, as raw material to fuel the creative spirit. An example of this would be the reason why someone would choose a particular profession. For money? No, because the economy is based on energy and the amount that would make you monetarily rich would also make you physically dead. For fame? No, since our media is not driven by the same motivations as they once were. The real goal is to learn, to gain knowledge. If that knowledge can be used to benefit someone at some point in time, so much the better.

        Our archaeologists recently uncovered ruins on the southwestern bank of Lake Michigan and found some puzzling documents. Many of them related to graft (or grafts) in "City Hall". Archeologists suspect that in that time all politicians owned extensive orchards and were very heavily into agriculture. Another document related to a scandal involving a "Watergate". They think it may have been a scandal involving a government official and a hydroelectric project.

        Our political system has changed as well. In the past they had to elect dictocrats that would make decisions based on the politicians self-interests rather than that of the people they were supposed to serve. And to add insult to injury they would make vast expenditures and live like royalty at the expense of the people who elected them. That doesn't happen now. You don't have to work just to survive. In fact, a lot are on a 10 to 15 hour work week. We make our own decisions through direct referendum. Actually it's a vast supercomputer complex that amounts to a large vote-tallying machine. In the past citizens would speak derisively of "the government". Now <u>we are</u> the government. There's no President, no House, no Senate. There are no middlemen. Anybody can bring up an issue for a D&D. That means debate and decision. The whole nation votes on it and the tally is computed in less than fifteen minutes. They had the technology to do this 500 years ago but it took a fundamental change in the value system and superior education to finally get it implemented. There are other political systems in use on the planet, and there is no attempt to force them to use ours. It's better to teach by example. The prime directive is used on this planet as well as others. perhaps I should explain what tthe "Prime Directive" is. Basically it forbids an advanced culture to interfere with a society which is technically or sociologically more primitive and less developed than it is. It seems odd that it took interstellar travel to teach us how to conduct ourselves on our own world. You see, the attitude that we once had was that we had an obligation to "fix" the universe. And part of that was improving the standard of living in that particular culture. One particular civilization died due to this. That culture was "improved" to their detriment. They had not advanced far enough to cope with the technology and philosophy which we had disseminated. In short, within a few years they had invented the electric toothbrush, and a year after that a method of blowing up the planet. That method proved to be all too effective.

        Our criminal justice system has changed as well. In the past institutions resembling storage pens (they must have been pens; after all, they called them **pen**itentiaries) were used to store those who had committed crimes against the populace. Actually these were just huge human meatlockers. In that time behavior modification was available but was not used due to the low esteem that was placed upon life. Life, in effect, was cheap. Now life is the most precious commodity. In the recent past, before the widespread use of free energy, felons were deported to a remote location in the world. Now with the adoption of the prime directive on this planet, that no longer is feasible. A twofold approach is used. The felon is given the choice of rehabilitation through behavior modification or a rather long detention period. That detention deserves some explanation. A scalar gravity field is generated in what appears to be a waiting room on the inside

The scalar gravity field induces relativistic effects, slows down the passage of time inside the room. For all intents and purposes it's about a hairs' breadth away from becoming a stasis field. Externally it appears to be a shiny cigar-shaped pod suspended in a vacuum chamber with shiny walls. Actually the whole assembly is a large thermos bottle that prevents extraneous heat from leaking into the pod. The walls of the chamber also contain a three-ply tesla shield configuration to cut down on radiation leakage. You see, since time is slowed down in the pod (as much as a factor of one million) all radiation is tremendously blueshifted in relationship to the pod. For example, incoming infrared which would have a frequency of about 10^{12} Hz., would be blueshifted to 10^{18} Hz. This would give those inside the pod a serious case of radiation poisoning. So, the shield is used to create an orthogonal rotation of the wave structure of the incoming radiation and eliminate the electromagnetic effects. Each 'waiting room' seats 1000. Ten are used for the entire nation. That's all that's needed. Usually they choose rehabilitation, which includes behavior modification to prevent a recurrence of the act. We never force rehabilitation on an individual because doing it would engender negative repercussions on society. So far there's been a 99.99% success rate. Perhaps later the transporter will be used to beam felons into some point in the future (that is, the ones who chose detention rather than rehabilitation).

    There are those who contend that when a particular civilization develops time travel, that society is doomed. At least that's what the sociologists say. Personally I think that the prospect of travelling in time is exciting, something to look forward to. It seems ironic that the only way to see if the society is destroyed by the proliferation of time travel is to use a time machine to go into the future and find out if that is what happens after all. There are several schools of thought that have existed through the centuries. One maintains that if you interact with the past you will change the past. Another one thinks that going into the past will enable the individual to participate in history without changing it. Yet another says that tripping off into the past will change only your own personal time-line and leave everybody else alone. In so many words, there exists an infinite number of universes. Each quantum of time fissions into a number of separate entities. Each entity represents a choice that has occurred in the universe. So that each choice exists in its own right and in its own universe. If you change something, the other universe that you have left still exists, it's just that you have permanently severed your contact with it. Some enterprising individuals are looking for a way to travel sideways in time to view what might have been, or what might have existed in our universe. If they're successful, then it would be possible for them to travel to the past, completely screw it up, and come back to the present to view the results. Then if they don't like it they can move sideways in time and come back to this universe. Or choose the best of all possible worlds and stay there. Anyway, it took the development of free energy and gravity control to develop our technology to the point to consider time travel. It could have been done in the past but there were just too many distractions. War, strikes, greed and every imaginable strife possible in quick succession destroyed the bright promise of hope in that time. Of course, all that is behind us now, and we are on the brink of the imponderable.

    Another possibility that this new technology has opened up is the prospect of altering environments on a planetary scale. Planetary engineering has risen from the realm of creative visualization to reality. It is true that you can use free energy to produce the most destructive bomb in the history of mankind... But it's also true that with shaped charges, entire planetary systems can be transformed for the better.

    Of course when you become the guardian of a technology which initiates thermonuclear ignition of entire suns, or can create white or black holes you also must accept the responsibility associated with an accident involving that specific application. It occurred in a prototype fusion plant located in Antarctica. Less than five watts of gravitational energy was used as a neg. grav. containment field in the reactor. The plasma fireball actually was a miniature neutron star with nuclear fusion occurring on a thin skin on the surface of the fireball. It took very little fuel to keep the reaction going in relation to the tremendous outpouring of energy from the miniature star. Then the unthinkable happened. A radiation-sensitive component that had been in all three backup systems failed. The power feed to the containment field rose drastically. The recording equipment that monitored the containment field peaked out at 100KW before it broke down. They had created a black hole, the first collapstar to be artificially generated. The bang was heard all the way to Rio, not to mention the earthquakes that followed. When the crew arrived at the scene all they saw was a crater 100 meters in diameter with a hole about a centimeter across running straight down into the rock. Then they realized what had happened. Eventually it leaked out to the rest of the world, which brought almost universal condemnation for the project. After the furor had died down they found that the black hole that had ignominiously been created was unstable and blew out after it penetrated the crust. That was a close one. It was also the last time they experimented with that design. I wonder why? It was later on when they found that the neutrino flux from the reactors using gravitational containment was not proportional to the energy output. So they used dud material for fuel and got nearly the same results that they expected. It turned out that the culprit was a direct transformation of gravitational energy to electromagnetic, due to scalar wave pair-coupling. Surprise!

Another change dealt with robotics. At first crude machines were constructed that could make use of the new form(s) of energy. Every household had an R2 model for a while. Later androids that had a somewhat human appearance were created. And then the real energy hogs were created. I use the term 'energy hogs' lightly. The only reason that robotics had been stuck in a self-limited paradigm is due to our concept of energy use.

The servant which is a projection of the central computer system in the house resembles a human; in fact he could even pass for human walking down a street! But he/it is merely a scalar-wave hologram which is projected by the machine. The projection itself, which is a standing wave, takes about 50 KWH to keep it energized. The refresh rate is dependant on the specific function that the projection is performing. You have to make sure that the power level is high enough, too. If the field is too 'soft' then the E.M. side of the triplex wave can become quenched if it comes in contact with something conductive. You should have seen what happened when the servant picked up the silverware. Let's just say that I had the fourth of July twice this year. After the incident with the rec. room and the 'droid I don't think that those guests will be over for a **looong** time!

The same process is used to automate manufacturing. Since anybody can make as many duplicates as he or she sees fit to make, all that's required in the way of manufacturing is to create high-quality originals that are fit to duplicate. That means very little in the way of production runs, but the runs are still required. The finished goods are assembled by taking a computer-generated triplex hologram and pouring raw energy into it. The finished product is then scanned, packing box and all, and then is offered to the consumer. Usually people see the holo-ad and then decide if they want it or not. Then the item is charged to their account and sent by transporter. Then you have the option to save it in the computer system and make a copy. Of course there is a sort of a blackmarket where a friend can supply you with a copy in return for a copy that they want. That's frowned upon. But it goes on anyway. That reminds me of those 22nd century clowns who copied themselves and got in trouble for cloning around. Specialized projections are used to move materials around and androids deal with the consumer.

Another culteral change dealt with our attitudes toward technology. When technology was used as a tool to survive and as a defense against a harsh environment, the society viewed it as utilitarian. Furniture, appliances, and vehicles had the spartan look. Now that we have time to include "frills", the artistic dimension has melded with the functional aspect of devices which we use. A few individuals who had a keen interest in conserving their energies devised a computer which actually generated paintings and poetry (not at the same time). Strange poetry. I have to admit that I've never heard haiku poetry about proctologists before. And I hope I never hear it again. That's what I consider a cruel and unusual punishment!

Well, it's time to set the wayback machine to return us to the present. I hope you enjoyed this short visit to a world that may be if free energy and gravity control are in widespread use. For if it is not to be used, I can guarantee that the future could be in a far gloomier state than this little scenario has described. For all our sakes, it would be in the best interest of all concerned if it were to be used.

Anything else would be fuelish.

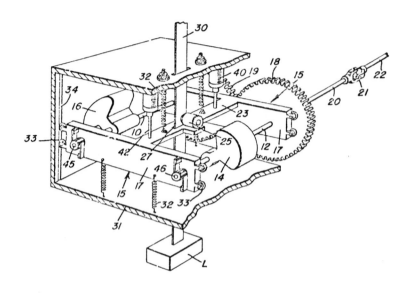

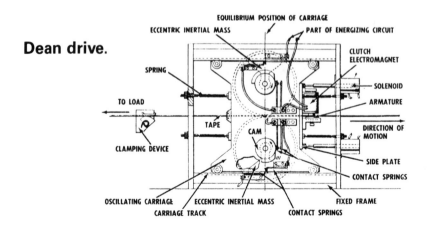

Dean drive.

Selected Patents & Their Principle Propulsive Forces

INVENTOR	ISSUED	MAIN PROPULSIVE FORCES
Llamozas	1953	impulse
Kellogg	1965	gyroscopic
di Bella	1968	Coriolis, centrifugal
Auweele	1970	impulse
Foster	1972	gyroscopic
Young	1971	gyroscopic, Coriolis, centrifugal
Lehberger	1975	centrifugal
Cook	1972	Coriolis, centrifugal, centripetal
Cook	1980	centrifugal

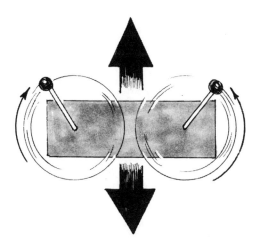

The heart of Dean's mechanism.

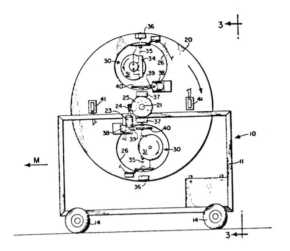

Foster drive.

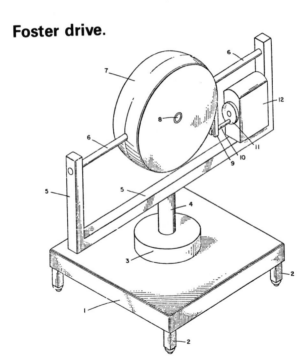

Kellogg drive.

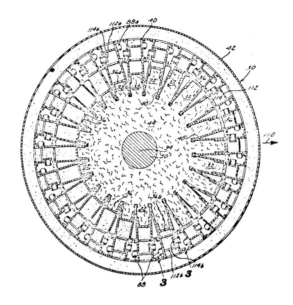

Matyas drive.

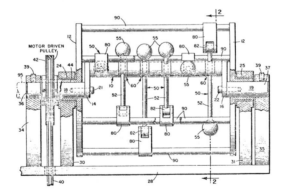

Novak drive.

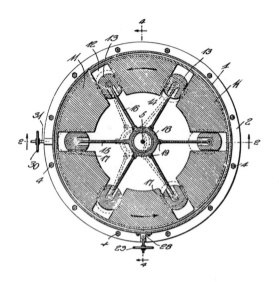

Laskowitz drive.

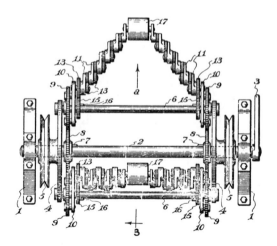

Nowlin drive.

Bibliography

How To Build a Flying Saucer
(And Other Proposals In Speculative Engineering)
T.B. Pawlicki (1981)
Prentice Hall, Inc., Englewood Cliffs, NJ.

How You Can Explore The Higher
Dimensions of Space and Time
(An Introduction To The New Science Of Hyperspace For Trekkies Of All Ages)
T. B. Pawlicki (1984)
Prentice Hall, Inc., Englewood Cliffs, NJ.

Ether Technology
Rho Sigma (1977)
Published by Rho Sigma
Printed for the publisher by CSA Printing & Bindery, Lakemont GA. 30552
You can get it through:
Tesla Book Company 1580 Magnolia Ave. Millbrae, CA. 94030

The Excalibur Briefing
Thomas E. Bearden (1980)
A Walnut Hill Book,
Strawberry Hill Press 2594 15th Avenue San Francisco, CA. 94127

Toward a New Electromagnetics Parts. 1,2,3 & 4
Thomas E. Bearden (1983)
Published by Tesla Book Company
1580 Magnolia Ave. Millbrae, CA. 94030

Solutions To Tesla's Secrets And The Soviet Tesla Weapons
Thomas E. Bearden (1983)
Tesla Book Company 1580 Magnolia Ave. Millbrae, CA. 94030

A Dual Ether Universe
Leonid Sokolow (1977)
Exposition Press Inc. 900 South Oyster Bay Road, Hicksville NY. 11801

Energy Unlimited No. 5
Energy Unlimited Rt. 4 Box 288 Los Lunas NM. 87031

The First International Symposium On Non-Conventional
Energy Technology-Proceedings

Stalking The Wild Pendulum
Itzhak Bentov (1977)
E. P. Dutton 2 Park Ave., New York, NY. 10016

Alternative (003)
Leslie Watkins (1977, 1978)
Avon Books, New York, NY. (originally a television series in Britain.)

The Philadelphia Experiment: Project Invisibility
Charles Berlitz & William L. Moore (1979)
Ballantine Books, New York, NY.

Harmonic 695, The UFO and Anti-Gravity
Bruce L. Cathie and Peter N. Temm (1971)
Quark Enterprises LTD 158 Shaw Rd., Oratia, Auckland, New Zealand
Telephone 818-4291

The Pulse of The Universe, Harmonic 288
Bruce Cathie (1977)
Quark Enterprises LTD 158 Shaw Rd., Oratia, Auckland, New Zealand

Harmonic 33
Bruce Cathie (1968)
Quark Enterprises LTD 158 Shaw Rd., Oratia, Auckland, New Zealand

The Bridge To Infinity
Bruce Cathie (1983)
Quark Enterprises LTD 158 Shaw Rd., Oratia, Auckland, New Zealand
Also available through Adventures Unlimited Press.

Introduction And Information Compendium
Volume 1- Antigravity And UfO's
Volume 2- Paranormal Phenomena
Volume 3- Energy
High Energy Electrostatics Research (or HEER) (1982)
P.O. Box 5286 Springfield, VA. 22150 R. A. Nelli, Director

The Hollow Earth
Raymond Bernard (1977)
Health Research, Mokelumne Hill, CA. 95245

Moongate: Supressed Findings Of The U. S. Space Program
The Nasa Military Cover-Up
William L. Brian II (1982)
Future Science Research Publishing Co.
P.O. Box 06392 Portland, Oregon 97206-0020

The Cosmic Conspiracy
Stan Deyo (1978)
West Australian Texas Trading
P.O. Box 71 Kalamunda, Western Australia 6076 or
William Collins PTY. LTD.
G.P.O. Box 476, Sydney, N.S.W., Australia 2001 or
Emissary Publications
P.O. Box 642 South Pasadena, CA. 91030

Flying Saucers - Serious Business
Frank Edwards (1966)
Bantam Books, Inc., 271 Madison Avenue, New York, NY. 10016

Somebody Else Is On The Moon
George Leonard (1976,1977)
Pocket Books, a Simon & Schuster Division of Gulf & Western Corporation
1230 Avenue of the Americas, New York, NY. 10020

The Roswell Incident
Charles Berlitz & William L. Moore (1980)
Grosset & Dunlap (A Filmways Company) New York, NY.

The House Of Lords UFO Debate
Lord Clancarty (Brinsley le Poer Trench) (Crown Copyright) (1979(?))
Open Head Press
2 Blenheim Crescent, London W11 1NN, Great Britain, in association with:
Pentacle Books
6 Perry Road, Bristol 1, Great Britain

Ramayana
Retold by William Buck (1976)
University Of California Press
Berkeley and Los Angeles, California

Mahabarata
Retold by William Buck (1973)
University Of California Press
Berkeley and Los Angeles, California

Mysteries Of The Unexplained
Reader's Digest (1982)
The Reader's Digest Association, Inc.
Pleasantville, New York/Montreal

New Horizons In Electric, Magnetic And Gravitational Field Theory
W. J. Hooper, BA, MA, PhD, (1974)
(President and Director of "Electrodynamic Gravity, Inc.)
Electrodynamic Gravity Inc.
543 Broad Blvd. Cuyahoga Falls, Ohio, 44

Future Physics And Anti-Gravity
William F. Hassel, PhD
MUFON Symposium Proceedings of 16,17 July 1977
4625 Stark Ave., Woodland Hills, CA. 91364

Energy Unlimited-A Case for Space
Arthur C. Aho (1968)
South Antelope Valley Publishing Company
Littlerock, California

Mysticism And The New Physics
Michael Talbot (1980)
Bantam Books, Inc.
666 5th Ave. New York, NY. 10103

No Earthly Explanation
John Wallace Spencer (1974)
Phillips Publishing Company
23 Hampden Street, Springfield, Massachusetts 01103

Flying Saucers Have Landed
Desmond Leslie & George Adamski (1953)
The British Book Centre, Inc.
420 West 45th Street, New York 36, NY.

Inside The Space Ships
George Adamski (1955)
Abelard-Schuman, Inc.
404 Fourth Ave. New York 16, NY.

Clear Intent: The Government Coverup Of The UFO Experience
Lawrence Fawcett, Barry J. Greenwood (1984)
Prentice-Hall Inc.
Englewood Cliffs, New Jersey 07632

UFO's Past Present & Future
Robert Emenegger (1974)
Ballantine Books A Division Of Random House, Inc.
201 E. 50th Street, New York, NY. 10022

Beyond Earth: Man's Contact With UFO's
Ralph Blum with Judy Blum (1974)
Bantam Books, Inc.
666 5th Ave. New York, NY. 10103

Gods Of Aquarius (UFO's And The Transformation Of Man)
Brad Steiger (1976)
Harcourt Brace Jovanovich, Inc.
757 Third Ave. New York, NY. 10017

Somebody Else Is On The Moon
George Leonard (1976)
The David McKay Company, Inc.
750 Third Ave. New York, NY. 10017

One Hundred Thoudand Years Of Man's Unknown History
Robert Charroux (1963)
Berkley Publishing Corporation
200 Madison Ave. New York, NY. 10016

Our Mysterious Spaceship Moon
Don Wilson (1975)
Dell Publishing Co., Inc.
1 dag Hammarskjold Plaza New York, NY. 10017

Secrets of Our Spaceship Moon
Don Wilson (1979)
Dell Publishing Co., Inc.
1 dag Hammarskjold Plaza New York, NY. 10017

From Outer Space
Howard Menger (1959)
Pyramid Books Mail Order Dept.
9 Garden Street Moonachie, NJ. 07074

Messangers of Deception
Jacques Vallee (1979,1980)
Bantam Books, Inc.
666 5th Ave. New York, NY. 10103

Space Frontier
Dr. Wernher von Braun (1963,64,65,66,67,69)
Fawcett World Library
67 West 44th Street, New York, NY. 10036

Mysteries of Time and Space
Brad Steiger (1974)
Dell Publishing Co., Inc.
1 dag Hammarskjold Plaza New York, NY. 10017

The Total UFO Story
Milt Machlin (1979)
Dale Books, Inc.,
380 Lexington Ave., New York, NY. 10017

Challenge To Science: The UFO Enigma
Jacques & Janine Vallee (1966)
Ballantine Books A Division of Random House, Inc.
201 E. 50th Street, New York, NY. 10022

Profiles of The Future
Arthur C. Clarke (1958,59,60,62)
Harper & Row, Publishers, Inc.
49 East 33rd Street, New York 16, NY.

Rockets, Missiles, And Men In Space
Willey Ley (1944,45,47,48,49,51,52,57,58,61,68)
The Viking Press Inc.
625 Madison Ave. New York, NY. 10022

Harmonic 33
Captain Bruce Cathie (1968)
A.H. & A.W. Reed Ltd.
65-67 Taranaki Street, Wellington, New Zealand

UFO's And Anti-Gravity: Contact With Earth (Harmonic 695)
Bruce L. Cathie And Peter N. Tamm (1971)
A Walnut Hill Book Strawberry Hill Press
616 4th Ave. San Francisco, California 94121

We Are Not The First
Andrew Tomas (1971)
Bantam Books, Inc.
666 Fifth Ave. New York, NY. 10019

Spacecraft Propulsion: New Methods
Hannes Alfven
Science April 14, 1972 pp. 167-168

Sould The Laws Of Gravitation Be Reconsidered?
Maurice F. G. Allasi
Aero/Space Engineering, Sept., 1959, pp. 46-52
Oct., 1959 pp. 51-55, Nov., 1959 pp. 55

The Future Of Aeronautics-Dreams and Realities
John E. Allen
The Aeronautical Journal, Sept., 1971 pp. 587-607

Soviet Efforts Should Be Closely Watched (Concerning Gravity Research)
William S. Beller
Missiles and Rockets, Sept. 11, 1961 pp. 27

Consultant's Report Overrides Dean Space Drive
William S. Beller
Missiles and Rockets Jun. 12, 1961, pp. 24

The Electric Field Rocket
H. C. Dudley
Analog, Nov., 1960

Gravitational Machines
F. J. Dyson
Gravity Research Foundation Essay, New Boston, New Hampshire, 1962
See Also:
Interstellar Communications
W. A. Benjamin, N.Y., 1963, pp. 115

Provocative Study Question's Einstein's Theory
Ryan Aeronautical Magazine. Outdated.

Breakthrough Foreseen in Early '70's
Missiles and Rockets, Feb. 15, 1965

Kinetic Gravity
Charles F. Brush
Science, Mar. 10, 1911, pp. 381-386

Impact/Impulse Drive
Harry W. Bull
American Journal Rocket Society No. 29, Sept. 1934, p 7-8

Mysterious New Aircraft powered by Reaction Motor
Harry W. Bull
Popular Science Monthly, Jan., 1935, pp. 27

(Letter) Construction details of Wilbur Smith's magnetic sink coil.
Analog, Dec. 1971, pp. 172

Final Report on the Dean Drive
John W. Campbell
Analog, Dec., 1960, pp. 4-6

Instrumentation for the Dean Drive
John W. Campbell
Analog, Nov. 1960, p. 95-96

Report on the Dean Drive
John W. Campbell
Analog, Sept. 1960 pp. 4-6

The Scientific Lynch Law
John W. Campbell
Analog, Oct., 1961 pp. 4

The Size of the Solar System
John W. Campbell
Analog, Jun., 1960 pp. 176

The Space Drive Problem
John W. Campbell
Analog, Jun., 1960 pp. 83

The Ultrafeeble Interactions
John W. Campbell
Analog, Dec. 1959 pp. 160

You Must Agree With Me
John W. Campbell
Analog, May 1960, pp. 177

Electrogravitics-What is-Or Might Be
A. V. Cleaver
Interplanetary Soc. J. Brit. Vol. 16 pp. 84-94, 1957

Is the Rocket the Only Answer
A. B. Cleaver
Interplanetary Soc. J. Brit., June 1947, pp. 127

How to Make a Flying Saucer
William D. Clendenon
Flying Saucers, Jun., 1964, pp. 36-47

Brass Tacks (Letter on solid state space drives.)
James E. Cox
Analog, Aug., 1968 pp. 174-175

Volume 1, No. 1&2, Jan.-Jun., 1969
James E. Cox (Editor)
Journal of Space Drive Research and Development (JOSDRAD)

Where the Reader Has His Say (Four types of space drives.)
James E. Cox
Flying Saucers, Feb. 1968 pp. 43-46

Indirect Physical Evidence
Roy Craig (1969)
Scientific Study of Unidentified Flying Objects
Bantam Books, N.Y., Jan., 1969 pp. 97-115

Antigravity Craft- Nature's Antigravity Devices
Guy J. Cyr
Sacred Heart Rectory
321 S. Broadway, Lawrence, Mass. 01843

The Fourth Law of Motion
William O. Davis
Analog, May 1962 pp. 83-104

Victory & Stine, Some Aspects of Certain Transient Phenomena
American Physical Society Bulletin (Abst.), Apr., 1962

Brass Tacks
Norman L. Dean
Analog, Jan., 1964 & May, 1963

Space Drive Rebuttal (Letter Dept.)
Norman L. Dean
Missiles and Rockets, Jun. 26, 1961 p. 6

Interstellar Transport
Freeman J. Dyson
Physics Today, Oct., 1968 pp. 41-45

Satellite Loses Weight
Frank Edwards
Strange World

Tomorrows Physics (1966) and The Momentor (1970)
John W. Ecklin
2721 S. June St.., Arlington, Va. 22202

Space Propulsion by Magnetic Field Interaction
J. F. Engelberger
Spacecraft J., pp. 347-349, 1964

A Self Propelling Mechanism
Lewis Epstein
The Physics Teacher, Vol. 8, pp. 332, Sept., 1970

Flying Saucers, Propulsion and Relativity
Gordon H. Evans
Fate, pp. 67-75

Major de Seversky's Ion Propelled Aircraft
Popular Science, Aug., 1964, pp. 58-64

Inertial Drive
Arthur Farall
Product Engineering, March 14, 1966, pp. 63

On the Possible Relation of Gravity to Electricity
Michael Faraday
Brittanica Great Books, Vol. 45, pp. 670-673

Force Field Shows Propulsion Promise
Missiles and Rockets
Jul. 11, 1960, pp. 27

Guidlines to Anti-Gravity
Robert L. Forward
American Journal of Physics 31, 1963, pp. 166-170

Particles That Travel Faster Than Light
Gerald Feinberg
Scientific American Feb., 1970, Vol. 222, No. 2, pp. 68

Zero Thrust Velocity Vector Control For Interstellar Probes:
Lorentx Force Navigation and Control
Robert L. Forward
AIAA Journal, Vol. 2, No. 5, May, 1964, pp. 855

Solar Sailing
R. L. Garwin
Jet Propulsion, 28, 1958, pp. 188-190

Otis T. Carr and the OTC-X1
Richard Gehman
True, Jan. 1961

Electrogravitic Propulsion
Lucien A. A.
Interavia, Vol. XI, No. 12, 1956

Elevators and Levitators
Cedric Giles
Journal of the American Rocket Society
Dec., 1946, No. 68, pp. 34-39

Essays on Gravity
Gravity Research Foundation
New Boston, New Hampshire

The Electromagnetic Cases
Richard Hall
UfO Evidence
Nicap Publication, 1536 Conneticut Ave., N.W. Washington, D.C. 20036

The General Limits of Space Travel
S. von Hoerner
Science, 137, pp. 18-23, 1962

Towards Flight Without Stress or Strain...or Weight
Intel, Washington, D.C.
Interavia, Vol. XI- No. 5, 1956, pp. 373

An Explanation of the Operating Principles of the Entropy Space Drive
Robert Jones
Space World, Jan., 1965, pp. 48

The Jetless Drive
Robert Jones
Amateur Rocketeer, Feb., 1964, pp. 16-19

Some Preliminary Evidence of a Mechanically Producible Directional Force Field
Robert Jones
Journal of Space Drive Research and Development (JOSDRAD)
Jan.-Mar., 1969, pp. 10-11

Abstract of above report.
Robert Jones
American Journal of Physics, Nov., 1969

Artist's conception of a vertical rising, disc-shaped aircraft which could result from a project under development for the U.S. Air Force by Avro Ltd., Canada *(Official U.S. Air Force photo)*.

Towards Flight
without Stress or Strain... or Weight

BY INTEL, WASHINGTON, D.C.

The following article is by an American journalist who has long taken a keen interest in questions of theoretical physics and has been recommended to the Editors as having close connections with scientific circles in the United States. The subject is one of immediate interest, and Interavia would welcome further comment from initiated sources.
— *Editors.*

Washington, D.C. — March 23, 1956 : Electro-gravitics research, seeking the source of gravity and its control, has reached a stage where profound implications for the entire human race begin to emerge. Perhaps the most startling and immediate implications of all involve aircraft, guided missiles — atmospheric and free space flight of all kinds.

If only one of several lines of research achieve their goal — and it now seems certain that this must occur — gravitational acceleration as a structural, aerodynamic and medical problem will simply cease to exist. So will the task of providing combustible fuels in massive volume in order to escape the earth's gravitic pull — now probably the shell. Some of the companies involved in this phase include Lear Inc., Gluhareff Helicopter and Airplane Corp., The Glenn L. Martin Co., Sperry-Rand Corp., Bell Aircraft, Clarke Electronics Laboratories, the U.S. General Electric Company.

The concept of weightlessness in conventional materials which are normally heavy, like steel, aluminium, barium, etc., is difficult enough, but some theories, so far borne out empirically in the laboratory, postulate that not only can they be made weightless, but they can in fact be given a negative weight. That is : the force of gravity will be repulsive to them and they will—new sciences breed new words and new meanings for old ones—loft away contra-gravitationally.

laboratory. Disc airfoils two feet in diameter and incorporating a variation of the simple two-plate electrical condenser charged with fifty kilovolts and a total continuous energy input of fifty watts have achieved a speed of seventeen feet per second in a circular air course twenty feet in diameter. More lately these discs have been increased in diameter to three feet and run in a fifty-foot diameter air course under a charge of a hundred and fifty kilovolts with results so impressive as to be highly classified. Variations of this work done under a vacuum have produced much greater efficiencies that can only be described as startling. Work is now under way developing a flame-jet generator to supply power up to fifteen million volts.

biggest headache facing today's would-be "space men".

And towards the long-term progress of mankind and man's civilization, a whole new concept of electro-physics is being levered out into the light of human knowledge.

There are gravity research projects in every major country of the world. A few are over 30 years old.[1] Most are much newer. Some are purely theoretical and seek the answer in Quantum, Relativity and Unified Field Theory mathematics — Institute for Advanced Study at Princeton, New Jersey; University of Indiana's School of Advanced Mathematical Studies; Purdue University Research Foundation; Goettingen and Hamburg Universities in Germany; as well as firms and Universities in France, Italy, Japan and elsewhere. The list, in fact, runs into the hundreds.

Some projects are mostly empirical, studying gravitic isotopes, electrical phenomena and the statistics of mass. Others combine both approaches in the study of matter in its super-cooled, super-conductive state, of jet electron streams, peculiar magnetic effects or the electrical mechanics of the atom's

In this particular line of research, the weights of some materials have already been cut as much as 30 % by " energizing " them. Security prevents disclosure of what precisely is meant by " energizing " or in which country this work is under way.

A localized gravitic field used as a ponderamotive force has been created in the

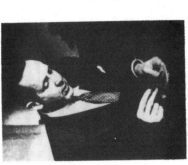

The American scientist Townsend T. Brown has been working on the problems of electrogravitics for more than thirty years. He is seen here demonstrating one of his laboratory instruments, a disc-shaped variant of the two-plate condenser.

Such a force raised exponentially to levels capable of pushing man-carrying vehicles through the air—or outer space—at ultra-high speeds is now the object of concerted effort in several countries. Once achieved it will eliminate most of the structural difficulties now encountered in the construction of high-speed aircraft. Importantly, the gravitic field that provides the basic propulsive force simultaneously reacts on all matter within that field's influence. The force is not a physical one acting initially at a specific point in the vehicle that needs then to be translated to all the other parts. It is an electro-gravitic field acting on all parts simultaneously.

Subject only to the so-far immutable laws of momentum, the vehicle would be able to change direction, accelerate to thousands of miles per hour, or stop. Changes in direction and speed of flight would be effected by merely altering the intensity, polarity and direction of the charge.

Man now uses the sledge-hammer approach to high-altitude, high-speed flight. In the still-short life-span of the turbo-jet airplane, he has had to increase power in the form of brute thrust some twenty times in order to achieve just a little more than twice the speed of the original jet plane. The cost in money

[1] Ultimately they go back to Einstein's general theory of relativity (1916), in which the law of gravitation was first mathematically formulated as a field theory (in contrast to Newton's " action-at-a-distance " concept). — Ed.

how little we do know; how vast is the area still awaiting research and discovery.

The most successful line of the electro-gravitics research so far reported is that carried on by Townsend T. Brown, an American who has been researching gravity for over thirty years. He is now conducting research projects in the U.S. and on the Continent. He postulates that there is between electricity and gravity a relationship parallel and/or similar to that which exists between electricity and magnetism. And as the coil is the usable link in the case of electro-magnetics, so is the condenser that link in the case of electro-gravitics. Years of successful empirical work have lent a great deal of credence to this hypothesis.

The detailed implications of man's conquest of gravity are innumerable. In road cars, trains and boats the headaches of transmission of power from the engine to wheels or propellers would simply cease to exist. Construction of bridges and big buildings would be greatly simplified by temporary induced weightlessness etc. Other facets of work now under way indicate the possibility of close controls over the growth of plant life; new therapeutic techniques; permanent fuel-less heating units for homes and industrial establishments; new sources of industrial power; new manufacturing techniques; a whole new field of chemistry. The list is endless... and growing.

In the field of international affairs, other

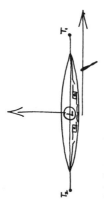

Author's sketch of a supersonic space ship roughly 50 ft. in diameter, whose lift and propulsion are produced by electrogravitic forces. The vehicle is supported by a "lofting cake" L consisting of "gravitic isotopes" of negative weight, and is moved in the horizontal plane by propulsion elements T_1 and T_2.

How soon all this comes about is directly proportional to the amount of effort that is put into it. Surprisingly, those countries normally expected to be leaders in such an advanced field are so far only fooling around. Great Britain, with her Ministry of Supply and the National Physical Laboratory, apparently has never seriously considered that the attempt to overcome and control gravity was worth practical effort and is now scurrying around trying to find out what it's all about. The U.S. Department of Defense has consistently considered gravity in the realm of basic theory and has so far only put token amounts of money into research on it. The French, apparently a little more open-minded about such things, have initiated a number of projects, but even these are still on pretty much of a small scale. The same is true throughout most of the

Townsend Brown's free-flying condenser. If the two arc-shaped electrodes (on the left and right rims) are placed under electrostatic charge, the disc will move, under the influence of interaction between electrical and gravitational fields, in the direction of the positive electrode. The higher the charge, the more marked will be the electrogravitic field. With a charge of several hundred kilovolts the condenser would reach speeds of several hundred miles per hour.

in reaching this point has been prodigious. The cost in highly-specialized man-hours is even greater. By his present methods man actually fights in direct combat the forces that resist his efforts. In conquering gravity he would be putting one of his most competent adversaries to work for him. Anti-gravitics is the method of the picklock rather than the sledge-hammer.

The communications possibilities of electro-gravitics, as the new science is called, confound the imagination. There are apparently in the ether an entirely new unsuspected family of electrical waves similar to electromagnetic radio waves in basic concept.

Electro-gravitic waves have been created and transmitted through concentric layers of the most efficient kinds of electro-magnetic and electro-static shielding without apparent loss of power in any way. There is evidence, but not yet proof, that these waves are not limited by the speed of light. Thus the new science seems to strike at the very foundations of Einsteinian Relativity Theory.

But rather than invalidating current basic concepts such as Relativity, the new knowledge of gravity will probably expand their scope, ramifications and general usefulness. It is this expansion of knowledge into the unknown that more and more emphasizes than electro-gravitics' military significance, what development of the science may do to raw materials values is perhaps most interesting to contemplate. Some materials are more prone to induced weightlessness than others. These are becoming known as *gravitic isotopes*. Some are already quite hard to find, but others are common and, for the moment, cheap. Since these ultimately may be the vital lofting materials required in the creation of contra-gravitational fields, their value might become extremely high with equivalent rearrangement of the wealth of national natural resources, balance of economic power and world geo-strategic concepts.

world. Most of the work is of a private venture kind, and much is being done in the studies of university professors and in the traditional lofts and basements of badly undercapitalized scientists.

But the word's afoot now. And both Government and private interest is growing and gathering momentum with logarithmic acceleration. The day may not be far off when man again confounds himself with his genius; then wonders why it took him so long to recognize the obvious.

Of course, there is always a possibility that the unexplained 3 % of UFO's, "Unidentified Flying Objects", as the U.S. Air Force calls "flying saucers", are in fact vehicles so propelled, developed already and undergoing proving flights—by whom... U.S., Britain... or Russia ? However, if this is so it's the best kept secret since the Manhattan project, for this reporter has spent over two years trying to chase down work on gravitics and has drawn from Government scientists and military experts the world over only the most blank of stares.

This is always the way of exploration into the unknown.

*

Author's diagram illustrating the electrogravitic field and the resulting propulsive force on a disc-shaped electrostatic condenser. The centre of the disc is of solid aluminium. The solid rimming on the sides is perspex, and in the trailing and leading edges (seen in the direction of motion) are wires separated from the aluminium core chiefly by air pockets. The wires act in a manner similar to the two plates of a simple electrical condenser and, when charged, produce a propulsive force. On reaching full charge a condenser normally loses its propulsive force, but in this configuration the air between the wires is also charged, so that in principle the charging process can be maintained as long as desired. As the disc also moves—from minus to plus—the charged air is left behind, and the condenser moves into new, uncharged air. Thus both charging process and propulsive force are continuous.

THE GRAVITICS SITUATION

<u>December 1956</u>
Gravity Rand Ltd.,
66 Sloane Street,
London S.W.1.

Theme of the science for 1956-1970:

SERENDIPITY

Einstein's view:-

"It may not be an unattainable hope that some day a clearer knowledge of the processes of gravitation may be reached; and the extreme generality and detachment of the relativity theory may be illuminated by the particular study of a precise mechanism".

CONTENTS

	Page
I - Engineering note on present frontiers of knowledge	3
II - Management note on the gravitics situation	10
III - Glossary	17
IV - References	20
V - Appendix	
Appendix I Summary of Townsend Brown's original specification for an apparatus for producing force or motion	22
Appendix II. Mozer's quantum mechanical approach to the existence of negative mass and its utilization in the construction of gravitationally neutralized bodies	30
Appendix III Gravity effects (Beams)	37
Appendix IV A link between Gravitation and nuclear energy (Deser and Arnowitt)	39
Appendix V Gravity/Heat interaction (Wickenden)	41
Appendix VI Weight-mass anomaly (Perl)	42

Thanks to the Gravity Research Foundation for Appendix II to VI

I Engineering note on present frontiers of knowledge

Gravitics is likely to follow a number of separate lines of development: the best known short term proposition is Townsend Brown's electrostatic propulsion by gravitators (details of which are to be found in the Appendix I). An extreme extrapolation of Brown's later rigs appears to suggest a Mach 3 interceptor type aircraft. Brown called this basically force and motion, but it does not appear to be the road to a gravitational shield or reflector. His is the brute force approach of concentrating high electrostatic charges along the leading edge of the periphery of a disk which yields propulsive effect. Brown originally maintained that his gravitators operate independently of all frames of reference and it is motion in the absolute sense - relative to the universe as a whole. There is however no evidence to support this. In the absence of any such evidence it is perhaps more convenient to think of Brown's disks as electrostatic propulsion which has its own niche in aviation. Electrostatic disks can provide lift without speed over a flat surface. This could be an important advance over all forms of airfoil which require induced flow; and lift - without-air flow is a development that deserves to be followed up in its own right, and one that for military purposes is already envisaged by the users as applicable to all three services. This point has been appreciated in the United States and a program in hand may now ensure that development of large sized disks will be continued. This is backed by the U.S. Government, but it is something that will be pursued on a small scale. This acceptance follows Brown's original suggestion embodied in Project Winterhaven. Winterhaven recommended that a major effort be concentrated on electrogravitics: based on the principle of his disks. The U.S. Government evaluated the disks wrongly, and misinterpreted the nature of the energy. This incorrect report was filed as an official assessment, and it

took some three years to correct the earlier misconception. That brings developments up to the fairly recent past and by that time it was realised that no effort on the lines of Winterhaven was practical, and that more modest aims should be substituted. These were re-written around a new report which is apparently based on newer thoughts and with some later patents not yet published - which form the basis of current U.S. policy. It is a matter of some controversy whether this research could be accelerated by more money but the impression in Gravity Rand is that the base of industry is perhaps more than adequately wide. Already companies are specializing in evolution of particular components of an electro-gravitics disk. This implies that the science is in the same state as the ICBM - namely that no new breakthroughs are needed, only intensive development engineering. This may be an optimistic reading of the situation: it is true that materials are now available for the condensers giving higher k figures than were postulated in Winterhaven as necessary, and all the ingredients necessary for the disks appear to be available. But industry is still some way from having an adequate power source, and possessing any practical experience of running such equipment.

The long term development of gravity shields, absorbers, and 'magic metals' appears at the moment however to be a basically different problem, and work on this is not being sponsored* so far as is known. The absorber or shield could be intrinsically a weapon of a great power, the limits of which are difficult to foresee. The power of the device to undermine the electrostatic forces holding the atom together is a destructive by-product of military significance. In unpublished work Gravity Rand has indicated the possible effect of such a device for demolition. The likelihood of such work being sponsored in small countries outside the U.S. is slight, since there is general lack of money and resources and in all such countries quick returns are essential.

* officially, that is

Many people hold that little or no progress can be made until the link in the Einstein unified field theory has been found. This is surely a somewhat defeatist view, because although no all-embracing explanation of the relationship between the extraordinary variety of high energy particles continually being uncovered is yet available much can be done to pin down the general nature of anti-gravity devices.

There are several promising approaches one of them is the search for negative mass, a second is to find a relationship between gravity and heat, and a third is to find the link between gravitation and the coupled particles. Taking the first of these: negative mass, the initial task is to prove the existence of negative mass, and appendix II outlines how it might be done this is Mozer's approach which is based on the Schroedinger time independent equation with the center of mass motion removed. As the paper shows this requires some 100 bev. which is beyond the power of existing particle accelerators: however the present Russian and American nuclear programs envisage 50 Tev, bevatrons within a few years and at the present rate of progress in the nuclear sciences it seems possible that the existence of negmass will be proved by this method of a Bragg analysis of the crystal structure or disproved. If negmass is established the precise part played by the subnuclear particles could be quickly determined. Working theories have been built up to explain how negative masses would be repelled by positive masses and pairs would accelerate gaining kinetic energy until they reach the speed of light and then assume the role of the high energy particles. It has been suggested by Ferrell that this might explain the role of neutrino, but this seems unlikely without some explanation of the spin ascribed to the neutrino. Yet the absence of rest mass or charge of the neutrino makes it especially intriguing. Certainly further study of the neutrino would be relevant to gravitational problems. If therefore the

aircraft industry regards anti gravity as part of its responsibilities it cannot escape the necessity of monitoring high energy physics of the neutrino. There are two aircraft companies definitely doing this but little or no evidence that most of the others know even what a neutrino is.

The relationship between electrical charges and gravitational forces however will depend on the right deductions being drawn from exceessively small anomalies? First clues to such small and hitherto unnoticed effects will come by study of the unified field theory: such effects may be observed in work on the gravithermals, and interacting effects of heat and gravity Here at least there is firmer evidence: materials are capable of temperature changes depending on gravity. This as Beams says (see Appendix III) is due to results from the alignment of the atoms Gravity tensions applied across the ends of a tube filled with electrolyte can produce heat or be used to furnish power The logical extension of this as an absorber of gravity in the form of a flat plate, and the gravitative flux acting on it (its atomic and molecular structure, its weight density and form are not at this stage clear)would lead to an increase in heat of the mass of its surface and subsurface particles

* * *

The third approach is to aim at discovering a connection between nuclear particles and the gravitational field This also returns to the need for interpreting microscopic relativistic phenomena at one extreme in terms of microscopic quantum mechanical phenomena at the other. Beaumont in suggesting a solution recalls how early theory established rough and ready assumptions of the characteristics of electron spin before the whole science of the atomic orbital was worked out. These were based on observation, and they were used with some effect at a time when data was needed Similar assumptions of complex spin might be used to link the microscopic to the macroscopic. At any rate there are some loose ends in complex spin to be tied up,

* See Appendix VI

-7-

and these could logically be sponsored with some expectation of results by companies wondering how to make a contribution

If a real spin or rotation is applied to a planar geoid the gravitational equipotentials can be made less convex, plane or concave. These have the effect of adjusting the intensity of the gravitational field at will, which is a requirement for the gravity absorber. Beaumont seemed doubtful whether external power would have to be applied to achieve this but it seems reasonable to suppose that power could be fed into the system to achieve a beneficial adjustment to the gravitational field, and conventional engineering methods could ensure that the weight of power input services would be more than offset by weightlessness from the spin inducer. The engineering details of this is naturally still in the realms of conjecture but at least it is something that could be worked out with laboratory rigs, and again the starting point is to make more accurate observations of small effects. The technique would be to accept any anomalies in nature, and from them to establish what would be needed to induce a spin artificially.

. . .

It has been argued that the scientific community faces a seemingly impossible task in attempting to alter gravity when the force is set up by a body as large as this planet and that to change it might demand a comparable force of similar planetary dimensions. It was scarcely surprising therefore that experience had shown that while it has been possible to observe the effects of gravity it resisted any form of control or manipulation But the time is fast approaching when for the first time it will be within the capability of engineers with bevatrons to work directly with particles that, it is increasingly accepted, contribute to the source of gravitation, and whilst that in itself may not lead to an absorber of gravity, it will at least throw more light on the

-9-

since it is already beginning to happen. There is not likely to be any sudden full explanation of the microcosm and macrocosm, but one strand after another joining them will be fashioned, as progress is made towards quantizing the Einstein theory.

-8-

sources of the power.

Another task is solution of outstanding equations to convert gravitational phenomena to nuclear energy The problem, still not yet solved, may support the Bondi-Hoyle theory that expansion of the universe represents energy continually annihilated instead of being carried to the boundaries of the universe. This energy loss manifests itself in the behaviour of the hyperon and K-particles which would, or might, form the link between the microcosm and macrocosm. Indeed Deseer and Arnowitt propose that the new particles are a direct link between gravitationally produced energy and nuclear energy. If this were so it would be the place to begin in the search for practical methods of gravity-manipulation. It would be realistic to assume that the K-particles are such a link. Then a possible approach might be to disregard objections which cannot be explained at this juncture until further unified field links are established As in the case of the spin and orbital theories, which were naive in the beginning, the technique might have to accept the apparent forces and make theory fit observation until more is known.

. . .

Some people feel that the chances of finding such a unified field theory to link gravity and electrodynamics are high yet think that the finding of a gravity shield is slight because of the size of the energy source, and because the chances of seeing unnoticed effect seem slender. Others feel the opposite and believe that a link between nuclear energy and gravitational energy may precede the link between the Einstein general relativistic and Quantum Theory disciplines. Some hope that both discoveries may come together while a few believe that a partial explanation of both may come about the same time, which will afford sufficient knowledge of gravitational fields to perfect an interim type of absorber using field links that are available. This latter seems the more likely

* See Appendix IV

-10-

II **Management Note on the Gravitics Situation**

The present anti-gravity situation as one of watching and waiting by the large aircraft prime contractors for lofting inventions or technological breakthroughs Clarence Birdseye in one of his last utterances thought that an insulator might be discovered by accident by someone working on a quite different problem and in 500 years gravity insulators would be commonplace. One might go further than Birdseye and say that principles of the insulator would, by then, be fundamental to human affairs it would be as basic to the society as the difference to-day between the weight of one metal and another But at the same time it would be wrong to infer from Birdseye's remark that a sudden isolated discovery will be the key to the science The hardware will come at a time when the industry is ready and waiting for it. It will arrive after a long period of getting accustomed to thinking in terms of weightlessness and naturally it will appear after the feasibility of achieving it in one form or another has been established in theory*

The aim of companies at this stage must therefore surely be to monitor the areas of progress in the world of high energy physics which seem likely to lead to establishment of the foundations of anti-gravity This means keeping a watchful eye on electrogravitics, magnetogravitics gravitics-isotopes and electrostatics in various forms for propulsion or levitation This is not at the present stage a very expensive business, and

* But this does not mean that harnessed forces will be necessarily fully understood at the outset

investment in laboratory man-hours is necessary only when a certain line of reasoning which may look promising comes to a dead-end for lack of experimental data, or only when it might be worth running some laboratory tests to bridge a chasm between one part of a theory and another, or in connecting two or more theories together. If this is right, anti gravity is in a state similar to nuclear propulsion after the NEPA findings, yet before the ANP project got under way. It will be remembered that was the period when the Atomic Energy Commission sponsored odd things here and there that needed doing. But it would be misleading to imply that hardware progress on electrostatic disks is presently so far along as nuclear propulsion was in that state represented by ANP. True the NEPA-men came to the conclusion that a nuclear-propelled aircraft of a kind could be built, but it would be only a curiosity. Even at the time of the Lexington and Whitman reports it was still some way from fruition. the aircraft would have been more than a curiosity but not competitive enough to be seriously considered.

It is not in doubt that work on anti-gravity is in the realm of the longer term future. One of the tests of virility of an industry is the extent to which it is so self confident of its position that it can afford to sponsor R&D which cannot promise a quick return. A closing of minds to anything except lines of development that will provide a quick return is a sign of either a straitlaced economy or of a pure lack of prescience,(or both).

Another consideration that will play its part in managerial decision is that major turning points in anti-gravity work are likely to prove far removed from the tools of the aircraft engineer. A key instrument for example that may determine the existence of negamass and establish posimass-negamass interaction is the super bevatron. It needs some 100 bev gammas on hydrogen to perform a Bragg analysis of the elementary particle structure by selective reflection to prove the existence of negamass. This value is double as much the new Russian bevatron under construction and it is 15 times as powerful as the highest particle accelerations in the Berkeley bevatron so far attained. Many people think that nothing much can be done until negamass has been observed. If industry were to adopt this approach it would have a long wait and a quick answer at the end. But the negamass-posimass theory can be further developed, and, in anticipation of its existence,means of using it in a gravitationally neutralized body could be worked out. This moreover is certainly not the only possible approach. a breakthrough may well come in the interaction between gravitative action and heat. Theory at the moment suggests that if gravity could produce heat the effect is limited at the moment to a narrow range* But the significant thing would be establishment of a principle

History may repeat itself thirty years ago, and even as recently as the German attempts to produce nuclear energy in the war, nobody would have guessed that power would be unlocked by an accident at the high end of the atomic table. All prophecies of atomic energy were concerned with how quickly means of fusion could be applied at the low end. In anti gravity work, and this

* See Appendix V

goes back to Birdseye, it may be an unrelated accident that will be the means of getting into the gravitational age. It is a prime responsibility of management to be aware of possible ways of using theory to accelerate such a process. In other words serendipity

It is a common thought in industry to look upon the nuclear experience as a precedent for gravity, and to argue that gravitics will similarly depend on the use of giant tools, beyond the capabilities of the air industry; and that companies will edge into the gravitational age on the coat tails of the Government as industry has done, or is doing, in nuclear physics. But this over looks the point that the two sciences are likely to be different in their investment. It will not need a place like Hanford or Savannah River to produce a gravity shield or insulator once the know-how has been established. As a piece of conceptual engineering the project is probably likely to be much more like a repetition of the turbine engine. It will be simple in its essence, but the detailed componentry will become progressively more complex to interpret in the form of a stable flying platform and even more intricate when it comes to applying the underlying principles to a flexibility of operating altitude ranging from low present flight speeds at one extreme to flight in a vacuum at the other this latter will be the extreme test of its powers Again the principle itself will function equally in a vacuum Townsend Brown s saucers could move in a vacuum readily enough but the supporting parts must also work in a vacuum In practice they tend to give trouble, just as gas turbine bits and pieces start giving trouble in proportion to the altitude gained in flight.

But one has to see this rise in complexity with performance and with altitude attainment in perspective:eventually the most advanced capability may be attained with the most extremely simple configurations. As is usual however in physics developments the shortest line of progress is a geodesic, which may in turn lead the propulsion trade into many roundabout paths as being the shortest distance between aims and achievement.

But aviation business is understandably interested in knowing precisely how to recognize early discoveries of significance, and this Gravity-Rand report is intended to try and outline some of the more promising lines. One suggestion frequently made is that propulsion and levitation may be only the last- though most important - of a series of others. some of which will have varying degrees of gravitic element in their constitution. It may be that the first practical application will be in the greater freedom of communications offered by the change in wave technique that it implies. A second application is to use the wave technique for anti-submarine detection, either airborne or seaborne. This would combine the width of horizon in search radar with the underwater precision of Magnetic Airborne Detection, and indeed it may have the range of scatter transmissions. Chance discoveries in the development of this equipment may lead to the fomulation of new laws which would define the relationship of gravity in terms of useable propulsion symbols. Exactly how this would happen nobody yet knows; and what industry and government can do at this stage is to explore all the possible applications simultaneously, putting pressure where results seem to warrant it.

In a paper of this kind it is not easy to discuss the details of the wave technique in communications, and the following are some of theories, briefly stated which require no mathematical training to understand, which it would be worth management keeping an eye on. In particular watch should be made of quantitative tests on lofting, and beneficiation of material. Even quite small beneficiation ratios are likely to be significant. There are some lofting claims being made of 20% and more, and the validity of these will have to be weighed carefully. Needless to say much higher ratios than this will have to be attained. New high-k techniques and extreme k materials are significant. High speeds in electrostatic propulsion of small discs will be worth keeping track of (by high speed one means hundreds of m.p.h.) and some of these results are beginning to filter through for general evaluation. Weight-mass anomalies, new oil-cooled cables, interesting megavolt gimmicks, novel forms of electrostatic augmentation with hydrocarbon and nonhydrocarbon fuels are indicative, new patents under the broadest headings of force and motion may have value, new electrostatic generator inventions could tip the scales and unusual ways of turning condensers inside-out, new angular propulsion ideas for barycentric control; and generally certain types of saucer configuration are valuable pointers to ways minds are working.

Then there is the personnel reaction to such developments. Managements are in the hands of their technical men, and they should beware of technical teams who are dogmatic at this stage. To assert electro-gravitics is nonsense is as unreal as to say it is practically extant. Management should be careful of men in their employ with a closed mind — or even partially closed mind on the subject.

This is a dangerous age when not only is anything possible, but it is possible quickly. a wise Frenchman once said "You have only to live long enough to see everything, and the reverse of everything" and that is true in dealing with very advanced high energy physics of this kind. Scientists are not politicians they can reverse themselves once with acclaim twice even with impunity. They may have to do so in the long road to attainment of this virtually perfect air vehicle. It is so easy to get bogged down with problems of the present;and whilst policy has to be made essentially with the present in mind — and in aviation a conservative policy always pays — it is management's task and duty to itself to look as far ahead as the best of its technicians in assessing the posture of the industry.

GLOSSARY

Gravithermels: alloys which may be heated or cooled by gravity waves. (Lover's definition)

Thermistors:
Electrads: } materials capable of being influenced by gravity.

Gravitator: a plurality of cell units connected in series; negative and positive electrodes with an interposed insulating member (Townsend Brown's definition).

Lofting: the action of levitation where gravity's force is more than overcome by electrostatic or other propulsion.

Beneficiation: the treatment of an alloy or substance to leave it with an improved mass-weight ratio.

Counterbary: this, apparently, is another name for lofting.

Barycentric control: the environment for regulation of lofting processes in a vehicle.

Modulation: the contribution to lofting conferred on a vehicle by treatment of the substance of its construction as distinct from that added to it by outside forces. Lofting is a synthesis of intrinsic and extrinsic agencies.

Absorber; insulator: these terms — there is no formal distinction between them as yet — are based on an analogy with electromagnetism. This is a questionable assumption since the similarity between electromagnetic and gravitational fields is valid only in some respects — such as both having electric and magnetic elements. But the difference in coupling strengths, noted by many experimenters, is fundamental to the science. Gravity moreover may turn out to be the only non-quantized field in nature, which would make it basically unique. The borrowing of terms from the field of electromagnetism is therefore only a temporary convenience. Lack of Cartesian representation makes this a baffling science for many people.

Negamass proposed mass that inherently has a negative charge.

Possimass mass the observed quantity - positively charged.

Shield a device which not only opposes gravity (such as an absorber) but also furnishes an essential path along which or through which, gravity can act. Thus whereas absorbers reflectors and insulators can provide a gravitationally neutralized body, a shield would enable a vehicle or sphere to 'fall away', in proportion to the quantity of shielding material.

Screening gravity screening was implied by Lanczos. It is the result of any combination of electric or magnetic fields in which one or both elements are not subject to varying permeability in matter.

Reflector a device consisting of material capable of generating buoyant forces which balance the force of attraction. The denser the material the greater the buoyancy force. When the density of the material equals the density of the medium the result will be gravitationally neutralized. A greater density of material assumes a lofting role.

Electrogravitics the application of modulating influences in an electrostatic propulsion system.

Magnetogravitics the influence of electromagnetic and meson fields in a reflector.

Bozum fields: these are defined as gravitational electro magnetic π and γ meson fields (Metric tensor).

Fermion fields these are electrons neutrinos muons nucleons and V particles (Spinors).

Gravitator cellular body: two or more gravitator cells connected in series within a body (Townsend Brown's definition)

F.A.E. Pirani and A. Schild, Physical Review 79 986 (1950).

Bergman, Penfield, Penfield, Schiller and Zatzkis, Physical Review 80, 81 (1950)

B.S. DeWitt, Physical Review 85, 653 (1952).

See, for example D. Bohm, Quantum Theory, New York, Prentice-Hall, Inc. (1951) Chapter 22.

B.S. DeWitt, Physical Review. 90, 357 (1953), and thesis (Harvard, 1950).

A. Pais, Proceedings of the Lorentz Kamerlingh Onnes Conference, Leyden, June 1953.

For the treatment of spinors in a unified field theory see W. Pauli, Annalen der Physik, 18, 337 (1933). See also B.S. DeWitt and C.M. DeWitt, Physical Review, 87, 116 (1952).

The Quantum Mechanical Electromagnetic Approach to Gravity F.L. Carter Essay on Gravity 1953.

On Negative mass in the Theory of Gravitation Prof. J.M. Luttinger Essay on Gravity 1951.

REFERENCES

Mackenzie Physical Review, 2 pp 321 43

Eotvos, Pekar and Fekete Annalen der Physik 68, (1922) pp 11-16.

Heyl, Paul R. Scientific Monthly, 47, (1938), p 115.

Austin, Thwing, Physical Review 5, (1897), pp 494 500

Shaw, Nature. (April 8, 1922) p 462 Proc. Roy. Soc., 102, (Oct. 6, 1922), p 46

Brush, Physical Review, 31, p 1113 (A)

Wold, Physical Review, 35, p 296 (abstract)

Majorana, Attidella Reale Accademie die Lincei, 28, (1919), pp 160, 221, 313, 416, 480 29 (1920) pp 23 90, 163, 235 Phil Mag. 39, (1920), p 288

Schneiderov, Science, (May 7, 1943), 97, sup. p 10.

Brush, Physical Review, 32 p 633 (abstract).

Lanczos, Science, 74, (Dec. 4, 1931), sup. p 10.

Eddington, Report on the Relativity Theory of Gravitation (1920) Fleetway Press, London

W.D. Fowler et al, Phys Rev. 93, 861, 1954.

R.L. Arnowitt and S. Deser, Phys Rev. 92 1061, 1953.

R.L. Arnowitt, Bull A.P.S. 94 798, 1954 S Deser, Phys. Rev. 93, 612, 1954.

N. Schein, D.M. Haskin and R G Glasser Phys. Rev 95 855, 1954

R.L. Arnowitt and S. Deser, unpublished Univ of California. Radiation Laboratory Report 1954

H. Bondi and T. Gold, Mon Not R Astr Soc 108 252 1948 F. Hoyle, Mon. Not R Astr Soc. 108, 372, 1948

B.S. DeWitt New Directions for Research in the Theory of Gravitation, Essay on Gravity 1953

C. H. Bondi. Cosmology, Cambridge University Press, 1952

APPENDIX I

SUMMARY OF TOWNSEND BROWN'S ORIGINAL PATENT SPECIFICATION

A Method of and an Apparatus or Machine for Producing Force or Motion.

This invention relates to a method of controlling gravitation and for deriving power therefrom, and to a method of producing linear force or motion. The method is fundamentally electrical.

The invention also relates to machines or apparatus requiring electrical energy that control or influence the gravitational field or the energy of gravitation; also to machines or apparatus requiring electrical energy that exhibit a linear force or motion which is believed to be independent of all frames of reference save that which is at rest relative to the universe taken as a whole, and said linear force or motion is furthermore believed to have no equal and opposite reaction that can be observed by any method commonly known and accepted by the physical science to date.

Such a machine has two major parts A and B. These parts may be composed of any material capable of being charged electrically. Mass A and mass B may be termed electrodes A and B respectively. Electrode A is charged negatively with respect to electrode B, or what is substantially the same, electrode B is charged positively with respect to electrode A, or what is usually the case, electrode A has an excess of electrons while electrode B has an excess of protons.

While charged in this manner the total force of A toward B is the sum of force g (due to the normal gravitational field), and

-23-

force e (due to the imposed electrical field) and force x (due to the resultant of the unbalanced gravitational forces caused by the electro negative charge or by the presence of an excess of electrons of electrode A and by the electro positive charge or by the presence of an excess of protons on electrode B)

By the cancellation of similar and opposing forces and by the addition of similar and allied forces the two electrodes taken collectively possess a force 2x in the direction of B This force 2x, shared by both electrodes, exists as a tendency of these electrodes to move or accelerate in the direction of the force, that is, toward B and B away from A. Moreover any machine or apparatus possessing electrodes A and B will exhibit such a lateral acceleration or motion if free to move.

In this Specification I have used terms as "gravitator cells" and "gravitator cellular body" which are words of my own coining in making reference to the particular type of cell I employ in the present invention. Wherever the construction involves the use of a pair of electrodes, separated by an insulating plate or member, such construction complies with the term gravitator cells, and when two or more gravitator cells are connected in series within a body, such will fall within the meaning of gravitator cellular body

The electrodes A and B are shown as having placed between them an insulating plate or member C of suitable material, such that the minimum number of electrons or ions may successfully penetrate it This constitutes a cellular gravitator consisting of one gravitator cell.

-24-

It will be understood that, the cells being spaced substantial distances apart the separation of adjacent positive and negative elements of separate cells is greater than the separation of the positive and negative elements of any cell and the materials of which the cells are formed being the more readily affected by the phenomena underlying my invention than the mere space between adjacent cells, any forces existing between positive and negative elements of adjacent cells can never become of sufficient magnitude to neutralize or balance the force created by the respective cells adjoining said spaces The uses to which such a motor, wheel, or rotor may be put are practically limitless as can be readily understood without further description. The structure may suitably be called a gravitator motor of cellular type.

In keeping with the purpose of my invention an apparatus may employ the electrodes A and B within a vacuum tube Electrons, ions, or thermions can migrate readily from A to B The construction may be appropriately termed an electronic ionic or thermionic gravitator as the case may be.

In certain of the last named types of gravitator units it is desirable or necessary to heat to incandescence the whole or a part of electrode A to obtain better emission of negative thermions or electrons or at least to be able to control that emission by variation in the temperature of said electrode A. Since such variations also influence the magnitude of the longitudinal force or acceleration exhibited by the tube it proves to be a very convenient method of varying this effect and of electrically controlling the motion of the tube The electrode A may be heated

-25-

to incandescence in any convenient way as by the ordinary methods utilizing electrical resistance or electrical induction

Moreover in certain types of the gravitator units, now being considered, it is advantageous or necessary also to conduct away from the anode or positive electrode B excessive heat that may be generated during the operation. Such cooling is effected externally by means of air or water cooled flanges that are in thermo connection with the anode, or it is effected internally by passing a stream of water, air, or other fluid through a hollow anode made especially for that purpose.

The gravitator motors may be supplied with the necessary electrical energy for the operation and resultant motion thereof from sources outside and independent of the motor itself. In such instances they constitute external or independently excited motors On the other hand the motors when capable of creating sufficient power to generate by any method whatsoever all the electrical energy required therein for the operation of said motors are distinguished by being internal or self excited Here, it will be understood that the energy created by the operation of the motor may at times be vastly in excess of the energy required to operate the motor. In some instances the ratio may be even as high as a million to one. Inasmuch as any suitable means for supplying the necessary electrical energy, and suitable conducting means for permitting the energy generated by the motor to exert the expected influence on the same may be readily supplied, it is now deemed necessary to illustrate details herein In said self-excited motors the energy necessary to overcome the friction or other resistance in the physical structure of the apparatus, and even to accelerate

-26-

the motors against such resistance, is believed to be derived solely from the gravitational field or the energy of gravitation. Furthermore, said acceleration in the self excited gravitator motor can be harnessed mechanically so as to produce usable energy or power, said usable energy or power, as aforesaid, being derived from or transferred by the apparatus solely from the energy of gravitation.

The gravitator motors function as a result of the mutual and unidirectional forces exerted by their charged electrodes The direction of these forces and the resultant motion thereby produced are usually toward the positive electrode. This movement is practically linear. It is this primary action with which I deal.

As has already been pointed out herein, there are two ways in which this primary action can accomplish mechanical work. First, by operating in a linear path as it does naturally, or second, by operating in a curved path Since the circle is the most easily applied of all the geometric figures, it follows that the rotary form is the important

There are three general rules to follow in the construction of such motors. First, the insulating sheets should be as thin as possible and yet have a relatively high puncture voltage It is advisable also to use paraffin saturated insulators on account of their high specific resistance. Second, the potential difference between any two metallic plates should be as high as possible and yet be safely under the minimum puncture voltage of the insulator Third, there should in most cases be as many plates as possible in order that the saturation voltage of the system might be raised well above the highest voltage limit upon which the motor is

-27-

operated Reference has previously been made to the fact that in the preferred embodiment of the invention herein disclosed the movement is towards the positive electrode However, it will be clear that motion may be had in a reverse direction determined by what I have just termed saturation voltage' by which is meant the efficiency peak or maximum of action for that particular type of motor: the theory, as I may describe it, being that as the voltage is increased the force or action increases to a maximum which represents the greatest action in a negative to positive direction If the voltage were increased beyond that maximum the action would decrease to zero and thence to the positive to negative direction

The rotary motor comprises broadly speaking, an assembly of a plurality of linear motors fastened to or bent around the circumference of a wheel In that case the wheel limits the action of the linear motors to a circle, and the wheel rotates in the manner of a fireworks pin wheel.

I declare that what I claim is

1. A method of producing force or motion which comprises the step of aggregating the predominating gravitational lateral or linear forces of positive and negative charges which are so co operatively related as to eliminate or practically eliminate the effect of the similar and opposing forces which said charges exert

2. A method of producing force or motion, in which a mechanical or structural part is associated with at least two electrodes or the like, of which the adjacent electrodes or the like have charges of differing characteristics, the resultant, predominating, uni directional gravitational force of said electrodes or the like

-28-

being utilized to produce linear force or motion of said part

3 A method according to Claim 1 or 2. in which the predominating force of the charges or electrodes is due to the normal gravitational field and the imposed electrical field.

4 A method according to Claim 1, 2 or 3. in which the electrodes or other elements bearing the charges are mounted, preferably rigidly, on a body or support adapted to move or exert force in the general direction of alignment of the electrodes or other charge-bearing elements

5 A machine or apparatus for producing force or motion, which includes at least two electrodes or like elements adapted to be differently charged, so relatively arranged that they produce a combined linear force or motion in the general direction of their alignment.

6. A machine according to Claim 5 in which the electrodes or like elements are mounted, preferably rigidly on a mechanical or structural part, whereby the predominating uni directional force obtained from the electrodes or the like is adapted to move said part or to oppose forces tending to move it counter to the direction in which it would be moved by the action of the electrodes or the like.

7. A machine according to Claim 5 or 6 in which the energy necessary for charging the electrodes or the like is obtained either from the electrodes themselves or from an independent source

8. A machine according to Claim 5 6 or 7 whose force action or motive power depends in part on the gravitational field or energy

-29-

of gravitation which is controlled or influenced by the action of the electrodes or the like

9. A machine according to any of Claims 5 to 8. in the form of a motor including a gravitator cell or a gravitator cellular body, substantially as described

10. A machine according to Claim 9, in which the gravitator cellular body or an assembly of the gravitator cells is mounted on a wheel-like support, whereby rotation of the latter may be effected, said cells being of electronic, ionic or thermionic type

11. A method of controlling or influencing the gravitational field or the energy of gravitation and for deriving energy or power therefrom comprising the use of at least two masses differently electrically charged, whereby the surrounding gravitational field is affected or distorted by the imposed electrical field surrounding said charged masses, resulting in a unidirectional force being exerted on the system of charged masses in the general direction of the alignment of the masses, which system when permitted to move in response to said force in the above mentioned direction derives and accumulates as the result of said movement usable energy or power from the energy of gravitation or the gravitational field which is so controlled, influenced, or distorted

12. The method of and the machine or apparatus for producing force or motion, by electrically controlling or influencing the gravitational field or energy of gravitation

-30-

APPENDIX II

A Quantum Mechanical Approach to the Existence of Negative Mass and Its Utilization in the Construction of Gravitationally Neutralized Bodies

Since the overwhelming majority of electrostatic quantum mechanical effects rely for their existence on an interplay of attractive and repulsive forces arising from two types of charge few if any fruitful results could come from a quantum mechanical investigation of gravity unless there should be two types of mass The first type, positive mass (hereafter denoted as posimass) retains all the properties attributed to ordinary mass, while the second type, negative mass (hereafter denoted as negamass) differs only in that its mass is an inherently negative quantity

By considering the quantum mechanical effects of the existence of these two types of mass, a fruitful theory of gravity will be developed Theory will explain why negamass has never been observed, and will offer a theoretical foundation to experimental methods of detecting the existence of negamass and utilizing it in the production of gravitationally neutralized bodies

To achieve these results recourse will be made to Schroedinger's time independant equation with the center of mass motion removed This equation is

$$-\hbar^2/2\mu \nabla^2 \psi + V\psi = E\psi$$

where all symbols represent the conventional quantum mechanical quantities Particular attention will be paid to the reduced mass $\mu = \frac{m_1 m_2}{m_1 + m_2}$ where m_1 and m_2 are the masses of the two interacting bodies.

One can approach the first obstacle that any theory of negamass faces, namely the explanation of why negamass has never been observed, by a consideration of how material bodies would be formed if a region of empty space were suddenly filled with many posimass and negamass quanta. To proceed along these lines, one must first understand the nature of the various possible quantum mechanical interactions of posimass and negamass.

Inserting the conventional gravitational interaction potential into Schroedinger's equation and solving for the wave function Ψ, yields the result that the probability of two posimass quanta being close together is greater than the probability of their being separated. Hence, there is said to be an attraction between pairs of posimass quanta. By a similar calculation it can be shown that while the potential form is the same, two negamass quanta repel each other. This arises from the fact that the reduced mass term in Schroedinger's equation is negative in this latter case. The type of negamass-posimass interaction is found to depend on the relative sizes of the masses of the interacting posimass and negamass quanta, being repulsive if the mass of the negamass quantum is greater in absolute value than the mass of the posimass quantum, and attractive in the opposite case. If the two masses are equal in absolute value, the reduced mass is infinite and Schroedinger's equation reduces to $(V-E)\Psi = 0$. Since the solution $\Psi = 0$ is uninteresting physically, it must be concluded that $V = E$, and, hence, there is no kinetic energy of relative motion. Thus, while there is an interaction potential between the equal mass posimass and negamass quanta, it results in no relative acceleration and thus, no mutual attraction or repulsion. While much could be said about the philosophical implications of the contradiction between this result and Newton's Second Law, such discussion is out of the scope of the present paper, and the author shall, instead, return with the above series of derivations to a consideration of the construction of material bodies in a region suddenly filled with many posimass and negamass quanta.

Because of the nature of the posimass posimass and negamass negamass interactions, the individual posimass quanta soon combine into small posimass spheres, while nothing has, as yet, united any negamass quanta. Since it is reasonable to assume that a posimass sphere weighs more than a negamass quantum in absolute value, it will attract negamass quanta and begin to absorb them. This absorption continues until the attraction between a sphere and the free negamass quanta becomes zero due to the reduced mass becoming infinite. The reduced mass becomes infinite when the sphere absorbs enough negamass quanta to make the algebraic sum of the masses of its component posimass and negamass equal to the negative of the mass of the next incoming negamass quantum. Thus, the theory predicts that all material bodies after absorbing as many negamass quanta as they can hold, weigh the same very small amount, regardless of size.

Since this prediction is in violent disagreement with experimental fact, one must conclude that the equilibrium arising as a result of the reduced mass becoming infinite has not yet been reached. That is, assuming that negamass exists at all, there are not enough negamass quanta present in the universe to allow posimass spheres to absorb all the negamass they can hold. One is thus able to explain the experimental fact that negamass has never been observed by deriving the above mechanism in which the smaller amounts of negamass that may be present in the universe are strongly absorbed by the greater amounts of posimass, producing bodies composed of both posimass and negamass but which have a net positive, variable, total mass.

Having thus explained why negamass has never been observed in the pure state, it is next desirable to derive an experimental test of the existence of negamass through considering the internal quantum mechanical problem of small amounts of negamass in larger posimass spheres. One is able to gain much physical insight into this problem by simplifying it to the qualitatively similar problem of one negamass quantum in the field of two posimass quanta that are fixed distance apart. Further simplification from three dimensions to one dimension and replacement of the posimass quanta potentials by square barriers, yields a solution in which the ground state energy E_0 of the negamass quantum in the field of one posimass quantum, is split into two energy levels in the field of the two posimass quanta. These two levels correspond to even and odd parity solutions of the wave equation where E_{even} lies higher and E_{odd} lower than E_0. The magnitudes of the differences $E_{even} - E_0$ and $E_0 - E_{odd}$ depend on the separation distance between the two posimass quanta, being zero for infinite separation and increasing as this separation distance is decreased.

Since the energy of a system involving negamass tends to a maximum in the most stable quantum mechanical configuration, the negamass quantum will normally be in state E_{even}. When the system is excited into state E_{odd} the negamass quantum will favor the situation in which the two posimass quanta are as far apart as possible, since E_{odd} increases with increasing separation distance between the two posimass quanta, and the system tends toward the highest energy state. Thus independent of and in addition to the attractive posimass posimass gravitational interaction there is a repulsive quantum mechanical exchange interaction between pairs of posimass quanta, when the system is in state E_{odd}. The result of these two oppositely directed interactions is that the two posimass quanta are in stable equilibrium at some separation distance. Since this equilibrium occurs between all posimass pairs in an elementary particle, a necessary consequence of the existence of negamass is that, when in the first excited state, elementary particles have a partial crystal structure.

This theoretical conclusion is capable of experimental verification by performing a Bragg analysis of the elementary particle crystal structure through shining high energy gamma rays on hydrogen. Part of the gamma ray energy will be utilized in lowering the system from energy E_{even} to E_{odd}, and if selective reflection is observed, it will constitute a striking verification of the existence of negamass. An order of magnitude calculation shows that, if the equilibrium distance between pairs of posimass quanta is one-one millionth the radius of an electron, 100 bev gamma rays will be required to perform this experiment.

Having discussed why negamass has never been observed, and having derived an experimental test of its existence, it is next desirable to develop an experimental method of utilizing negamass in the production of gravitationally neutralized bodies by further consideration of some ideas previously advanced. It has been pointed out that if a source of negamass is present, a posimass sphere continues to absorb negamass quanta until equilibrium is reached as a result of the reduced mass becoming infinite. Because the sphere thus produced is practically massless, and because the gravitational interaction between two bodies is proportional to the product of their respective masses, it follows that the sphere is practically unaffected by the presence of other bodies. And thus the problem of making gravitationally neutralized bodies is reduced to the problem of procuring a source of negamass quanta. This will be the next problem discussed.

The binding energy of a negamass quantum in a posimass sphere may be obtained as one of the eigenvalue solutions to Schroedinger's Equation. If the negamass quanta in a body are excited to energies in excess of this binding energy by shining sufficiently energetic gamma rays on the body these negamass quanta will be emitted and a negamass source will thus be obtained

To estimate the gamma ray energy required to free a negamass quantum from a posimass body, certain assumptions must be made concerning the size and mass of posimass and negamass quanta Since these quantities are extremely indefinite, and since the whole theory is at best qualitative, attempting to estimate the energy would be a senseless procedure Suffice it to say that because of the intimate sub elementary particle nature of the posimass-negamass interaction, it seems reasonable to assume that quite energetic gamma rays will be required to break this strong bond.

To briefly review what has been shown a quantum mechanical theory of negamass has been developed, based on the assumptions that gravitational interactions obey the laws of quantum mechanics and that all possible interactions of negamass and posimass with themselves and each other follow the well known inverse square law This theory explains the experimental fact that negamass has never been observed, and outlines plausible experimental methods of determining the existence of negamass and utilizing it in the construction of gravitationally neutralized bodies While these experimental methods may perhaps be out of the realm of practicality at the present, there is every reason to hope that they will be performable in the future. At that time, the plausibility of the existence of negamass and the theory behind the construction of gravitationally neutralized bodies from it will meet their final tests.

SUMMARY PARAGRAPH

A quantum mechanical theory of negative mass is developed, based on the assumptions that gravitational interactions obey the laws of quantum mechanics and that all possible interactions of negative and positive mass with themselves and each other follow the well known inverse square law. This theory explains the experimental fact that negative mass has never been observed, and outlines plausible experimental methods of determining the existence of negative mass and utilizing it in the construction of gravitationally neutralized bodies

Prof F. Mozer

APPENDIX III

GRAVITY EFFECTS

The order of magnitude of the heat given off by an alloy as a result of the separation by gravity tension can be reliably estimated Suppose we assume that an alloy of half tin and half lead completely fills a tube 5 meters long and 100 cm^2 cross section which is maintained accurately at a temperature 277° C At this temperature the alloy is liquid. Suppose next that the tube is raised from a horizontal plane into a vertical position. i.e. to a position where its length is parallel to the direction of gravity If then the alloy is free from convection as it would be if it is maintained at uniform temperature and if it is held in this position for several months, the percentage of tin at the bottom of the tube will decrease while the relative amount at the top will increase A simple calculation shows that the concentration of tin at the top is about one tenth of one percent greater than at the bottom and that approximately one calorie of heat is given off in the separation progress. If after several months the tube is again placed so that its length is in a horizontal plane the tin and lead will remix due to the thermal agitation of the atoms and heat is absorbed by the alloy.

Another interesting effect occurs when an electrolyte is subjected to gravity tension Suppose a five meter glass tube is filled with a water solution of say barium chloride and the electrical potential between its ends is measured first when the length of the tube is parallel to the horizontal and second, when its length is vertical. The difference in potential between the two ends is practically zero when the tube is horizontal and approximately eighty five microvolts when it is vertical This effect was discovered by Des Coudres in 1892 If a resistor is attached across the ends when the tube is vertical, heat of course is produced If the tube is maintained at constant temperature the voltage decreases with time and eventually vanishes The effect is believed to result from the fact that the positively charged barium ions settle faster than the lighter negatively charged chlorine ions as a result of gravity tension

In conclusion we have seen that gravity tension effects an alloy in such a way that it gives off heat This phenomenon results from the alignment of the atoms and from their separation by the gravitational field, the contribution of the latter being larger than that of the former. Also the gravity tension sets up a potential across the ends of a tube filled with an electrolyte and this potential when applied accross an external circuit may produce heat or drive an electric motor to furnish power Several other small thermal effects possibly may arise from gravity tension in addition to those discussed above but space is not available to consider them in this essay Also studies of the effect of gravitational fields and their equivalent centrifugal fields upon matter will no doubt be of great value in the future

J W Beams

APPENDIX IV

A LINK BETWEEN GRAVITATION AND NUCLEAR ENERGY
by Dr Stanley Deser
and
Dr Richard Arnowitt

Quantitatively we propose the following field equations

$$-kT_{\mu\nu} = R_{\mu\nu} + \tfrac{1}{2} R g_{\mu\nu} + C_{\mu\nu}(\phi\psi)$$

$$(\tfrac{1}{i}\delta^{\mu}\partial_{j\mu} + m + \lambda \sigma^{\mu\nu} K_{\mu\nu}(x))\psi = 0$$

with a similar equation for ϕ. In the above ψ represents the hyperon wave functions, and ϕ the K particle quantized field operators. The first three terms in the first equation are the usual structures in the Einstein General Relativity. The last term, $C_{\mu\nu}$ is the "creation" tensor which is to give us our conversion from gravitational to nuclear energy. It is like $T_{\mu\nu}$ in being an energy momentum term. In the second equation ∂j_μ represents the covariant derivative while δ^μ is a generalized Dirac matrix arranged so that the second equation is indeed covariant under the general group of coordinate transformations. The $\sigma^{\mu\nu} K_{\mu\nu}$ term will automatically include the higher hyperon levels. $C_{\mu\nu}$ is a functional of the hyperon and K field variables ψ and ϕ. As can be seen these equations are coupled in two ways: first the creation term $C_{\mu\nu}$ depends upon the field variables ψ and ϕ while the gravitational metric tensor $g_{\mu\nu}$ enters through the covariant derivative etc; λ is a new universal constant giving the scale of the level spacings of the hyperons. Rigorously speaking the field equations should be of course second quantized. For purposes of obtaining a workable first approximation it is probably adequate to take expectation values and solve the semi classical equations. The creation tensor $C_{\mu\nu}$ must be a bilinear integral of the ϕ and ψ fields and may have cross terms as well of the form $\int \phi \psi \psi (dx)$. These equations will indeed be difficult to solve but upon solution will give the distribution of created energy and hence lead eventually to the more practical issues desired.

APPENDIX V

GRAVITY/HEAT INTERACTION

Let us suppose that we have to investigate the question whether gravitative action alone upon some given substance or alloy can produce heat. We do not specify its texture density nor atomic structure. we assume simply the flux of gravitative action followed by an increase of heat in the alloy.

If we assume a small circular surface on the alloy then the gravitative flux on it may be expressed by Guass theorem and it is $4\pi M$, where M represents mass of all sub surface particles the question is, can this expression be transformed into heat. We will assume it can be. Now recalling the relativity law connecting mass and energy :—

$$M = m_o + \frac{T}{c^2} \quad (\text{by Einstein})$$

where:— T – Kinetic energy
m – Initial mass
c – Velocity of Light

we set $4\pi M = m_o + \frac{T}{c^2} = m_o + \frac{m_o v^2}{2c^2}$,

But $\frac{v^2}{c^2}$ is a proper fraction : hence $M = m_o + \frac{m_o}{2K}$

In the boundary case $V = C$, $M = m_o(1 + \frac{1}{K})$ for all other cases $4\pi M = m_o(\frac{K+1}{K}) K \neq 0$ Strictly M should be preceded by a conversion factor $\frac{1}{K}$ but if inserted. it does not alter results Thus if gravity could produce heat. the effect is limited to a narrow range, as this result shows.

It merits stress that in a gravitational field the flow lines lines of descent are Geodesics.

J.W Wickenden

APPENDIX VI

WEIGHT MASS ANOMALY

There is a great need for a precise experimental determination of the weight to mass ratio of protons or electrons. Since the ratio for a proton plus an electron is known already, the determination of the ratio for either particle is sufficient. The difficulty of a direct determination of the gravitational deflection of a charged particle in an experiment similar to the neutron or neutral atom experiment is due to electrical forces being much greater than gravitational forces. For example one electron five meters away from a second electron exerts as much force on that second electron as the gravitational field does. Thus stray electrons or ions which are always present on the walls of an apparatus can exert sufficient force to completely mask the gravitational force. Even if the surface charges are neglected image charges of the electron beam itself and self repulsion in the beam may obscure the gravitational deflection. An additional problem is the Earth's magnetic field. Electrons of even a few volts energy will feel a force due to the Earth's field a thousand billion times larger than the gravitational deflection. This last problem is avoided in a static measurement of the ratio such as a weighing of ionized matter. However this last method has the additional difficulty of requiring a high proportion of ionized to unionized matter in the sample being weighed. Of course all these problems can be resolved to some extent but it is questionable if an experiment of either of the above types can be designed in which all the adverse effects can simultaneously be sufficiently minimized. Probably a completely new type of experiment will have to be devised to measure the weight to mass ratio of the proton or electron. Such a measurement may detect a deviation from the law of constant weight to mass ratio. If such an anomaly can be shown to exist there is the possibility of finding a material which would be acted upon in an unusual manner in a gravitational field.

Martin L. Perl.

The Top Secret UFO Base Called Area 51

Area 51 is a part of the "Dreamland" complex where America's most advanced weaponry systems are believed to be under developement. The Area 51 portion of that vast desert facility is where it is claimed that the U. S. Government is now examining all UFOs that have either crash-landed or have been captured by a supposed elite unit of the military. These spacecrafts, and their occupants, are now said to be undergoing examination by top specialists in the medical, metallurgical, and propulsion fields.

The entire Dreamland complex is mainly underground to prevent any unauthorized observation from satellites, overflights, or individuals hiking in the surrounding mountains. It should be mentioned that this facility is guarded by a small army of current military and intelligence services personnel, as well as a reputed hand-picked force of ex-servicemen who served in the Navy's Seals, Army's Airborne or Special Forces, and the Air Force's Air Commandos. It is known that these guards have at their disposal armored vehicles, helicopters, mobile radar units, highly sensitive detectors on the ground, and more. There is

Nellis Bombing and Gunnery Range

From the book <u>Red Flag</u> by Michael Skinner, 1984, Presidio Air Power Series.

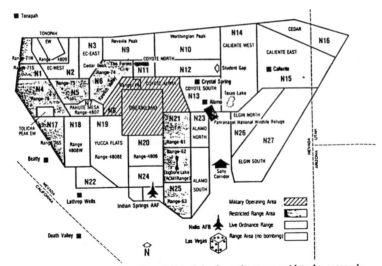

"Dreamland" includes Site 51 and Site S-4 where there are said to be research and operational facilities in above and below ground installations for flying discs. There are allegedly also environmental habitats for extraterrestrial intelligent beings. (Information obtained from various UFO researchers).

also a strong working relationship with the Lincoln County Sheriff's department, for any kind of assistance when called.

All U. S. Government vehicles that are authorized to enter this restricted area have on their front license plate a "CSC" tag. It has been speculated that "CSC" stands for "Central Space Center", but nobody who knows is talking about what that logo means.

Area 51 was first coded "Operation Snowbird" and it is believed that its main mission was to test fly captured UFOs. *Hangar 18* was a movie about a captured UFO that the U. S. Government was holding in "hangar 18". When the film was first shown to the public it caused a lot of concern at the alleged home of hangar 18, Wright-Patterson Air Force Base in Ohio. Supposedly a decision was made to find a new location for the recovered saucers at "Wright-Pat". In 1972 the most favorable site for selection was found in Nevada due to its remoteness from the general public, and was named Area 51. This was an ideal location because of two concealing mountain ranges in the dry lake bed area, Groom Lake, and the only road into the area was soon up-graded to a two-lane one suitable for heavy-duty trucks.

Directions to Area 51

From Las Vegas, Nevada, travel in a north-easterly direction on Interstate I-15 (heading for Salt Lake City, Utah) for a distance of about 22 miles, then turn left onto U.S. Hwy 93 for about 85 miles, then turn left again onto Nevada State Hwy 375 for about 14 miles, then left once more at the bottom of Hancock Summit onto a two-land gravel road that is well-maintained for about 13 miles, follow this until you reach the first guard post that is located on top of the Groom Mountain area. See map below.

Remember, the U. S. Government, and possibly extraterrestrial intelligences, do not want you there. Approach with caution, do not break the law. Dead UFO researchers are not terrifically useful!

Above: A 1968 satellite photo of the Groom Lake (dry)/Area 51 site, near Las Vegas, Nevada. **Left:** Bob Lazar, a physicist, claims that he worked at the S-4 site, also known as "Dreamland," which is located in Area 51 of the Nellis Bombing and Gunnery Range. **Below:** Lazar claims that top secret anti-gravity experiments with discoid craft take place within the various underground floors of the gigantic complex. Lazar claims to have seen a craft resembling this "flying saucer" within the facility. Is this a man-made or extraterrestrial craft?

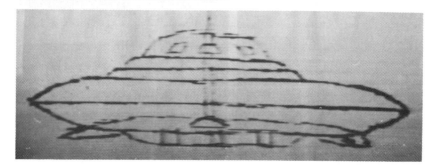

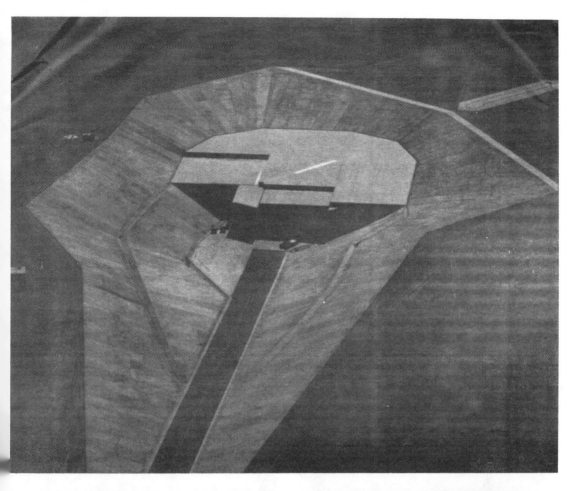

Above: An alleged photo of one of Lockheed Aircraft's covert operations center Helendale, California. It is apparently the entrance to an underground facility. **Bottom photos:** A sliding door to an underground facility and (right) a crane, at the McDonnell Douglas covert facility northeast of Llano, California.

THE ADVENTURES UNLIMITED CATALOG

Order from the following pages or write for our free 56 page catalog of unusual books & videos!

NEW BOOKS FROM ADVENTURES UNLIMITED

THE CASE FOR THE FACE
Scientists Examine the Evidence for Alien Artifacts on Mars
edited by Stanley McDaniel and Monica Rix Paxson
Mars Imagery by Mark Carlotto

The ultimate compendium of analyses of the Face on Mars and the other monuments in the Cydonia region. *The Case For the Face* unifies the research and opinions of a remarkably accomplished group of scientists, including a former NASA astronaut, a quantum physicist who is the chair of a space science program, leading meteor researchers, nine Ph.D.'s, the best-selling science author in Germany and more. The book includes: NASA research proving we're not the first intelligent race in this solar system; 120 amazing high resolution images never seen before by the general public; three separate doctoral statistical studies demonstrating the likelihood of artificial objects at the Cydonian site to be over 99%; and other definitive proof of life on Mars. Solid science presented in a readable, richly illustrated format. This book will also be featured in a Learning Channel special featuring Leonard Nimoy.
320 PAGES. 6X9 PAPERBACK. ILLUSTRATED. INDEX & BIBLIOGRAPHY. $17.95. CODE: CFF

FLYING SAUCERS OVER LOS ANGELES
The UFO Craze of the '50s
by DeWayne B. Johnson & Kenn Thomas
commentary by David Hatcher Childress

Beginning with a previously unpublished manuscript written in the early 1950s by DeWayne B. Johnson entitled "Flying Saucers Over Los Angeles," this book chronicles the earliest flying saucer flap beginning June 24, 1947. The book continues with other sightings into the late '50s, including many rare photos. It also presents one of the first analyses of the sociological and psychological dimensions of the UFO experience, from a vantage point of certainty that flying saucers are real—borne out by the actual news and witness accounts. Starting with such cases as the Roswell crash and the Maury Island incident, it continues to little-known sightings; this manuscript offers contemporaneous view of the earliest UFO excitement in America, unvarnished by the accumulated speculation of the last 46 years. A more detailed account of the many early sightings has never before been published. Additionally, the book contains a gazetteer of UFO sightings from 1947 up to the late-'50s.
256 PAGES. 6X9 PAPERBACK. ILLUSTRATED. BIBLIOGRAPHY & INDEX. $16.00. CODE: FSLA

HAARP
The Ultimate Weapon of the Conspiracy
by Jerry Smith

The HAARP project in Alaska is one of the most controversial projects ever undertaken by the U.S. Government. Jerry Smith gives us the history of the HAARP project and explains how it can be used as an awesome weapon of destruction. Smith exposes a covert military project and the web of conspiracies behind it. HAARP has many possible scientific and military applications, from raising a planetary defense shield to peering deep into the earth. Smith leads the reader down a trail of solid evidence into ever deeper and scarier conspiracy theories in an attempt to discover the "whos" and "whys" behind HAARP and their plans to rule the world. At best, HAARP is science out-of-control; at worst, HAARP could be the most dangerous device ever created, a futuristic technology that is everything from super-beam weapon to world-wide mind control device. The Star Wars future is now.
248 PAGES. 6X9 PAPERBACK. ILLUSTRATED. BIBLIOGRAPHY & INDEX. $14.95. CODE: HARP

THE HARMONIC CONQUEST OF SPACE
by Captain Bruce Cathie

A new, updated edition with additional material. Chapters include: Mathematics of the World Grid; the Harmonics of Hiroshima and Nagasaki; Harmonic Transmission and Receiving; the Link Between Human Brain Waves; the Cavity Resonance between the Earth; the Ionosphere and Gravity; Edgar Cayce—the Harmonics of the Subconscious; Stonehenge; the Harmonics of the Moon; the Pyramids of Mars; Nikola Tesla's Electric Car; the Robert Adams Pulsed Electric Motor Generator; Harmonic Clues to the Unified Field; and more. Also included in the book are tables showing the harmonic relations between the Earth's magnetic field, the speed of light, and anti-gravity/gravity acceleration at different points on the Earth's surface.
248 PAGES. .X9. TRADEPAPER. ILLUSTRATED. BIBLIOGRAPHY. $16.95. CODE: HCS

RETURN OF THE SERPENTS OF WISDOM
by Mark Amaru Pinkham

According to ancient records, the patriarchs and founders of the early civilizations in Egypt, India, China, Peru, Mesopotamia, Britain, and the Americas were colonized by the Serpents of Wisdom—spiritual masters associated with the serpent—who arrived in these lands after abandoning their beloved homelands and crossing great seas. While bearing names denoting snake or dragon (such as Naga, Lung, Djedhi, Amaru, Quetzalcoatl, Adder, etc.), these Serpents of Wisdom oversaw the construction of magnificent civilizations within which they and their descendants served as the priest kings and as the enlightened heads of mystery school traditions. The Return of the Serpents of Wisdom recounts the history of these "Serpents"—where they came from, why they came, the secret wisdom they disseminated, and why they are returning now.
332 PAGES. 6X9 PAPERBACK. ILLUSTRATED. REFERENCES. $16.95. CODE RSW

24 HOUR CREDIT CARD ORDERS—CALL: 815-253-6390 FAX: 815-253-6300

NEWLY RELEASED

LOST CONTINENTS & THE HOLLOW EARTH
I Remember Lemuria and the Shaver Mystery
by David Hatcher Childress & Richard Shaver

Lost Continents & the Hollow Earth is Childress' thorough examination of the early hollow earth stories of Richard Shaver and the fascination that lost continents and the hollow earth have had for the American public. Shaver's rare 1948 book *I Remember Lemuria* is reprinted in its entirety, and the book is packed with illustrations from Ray Palmer's *Amazing Stories* magazine of the 1940s. Palmer and Shaver told of tunnels running through the earth—tunnels inhabited by the Deros and Teros, humanoids from an ancient spacefaring race that had inhabited the earth, eventually going underground, hundreds of thousands of years ago. Childress discusses the famous hollow earth books and delves deep into whatever reality may be behind the stories of tunnels in the earth. Operation High Jump to Antarctica in 1947 and Admiral Byrd's bizarre statements, tunnel systems in South America and Tibet, the underground world of Agartha, UFOs coming from the South Pole, more.
344 PAGES. 6X9 PAPERBACK. ILLUSTRATED. $16.95. CODE: LCHE

INSIDE THE GEMSTONE FILE
Howard Hughes, Onassis & JFK
by Kenn Thomas & David Hatcher Childress

Steamshovel Press editor Thomas takes on the Gemstone File in this run-up and run-down of the most famous underground document ever circulated. Photocopied and distributed for over 20 years, the Gemstone File is the story of Bruce Roberts, the inventor of the synthetic ruby widely used in laser technology today, and his relationship with the Howard Hughes Company and ultimately with Aristotle Onassis, the Mafia, and the CIA. Hughes kidnapped and held a drugged-up prisoner for 10 years; Onassis and his role in the Kennedy Assassination; how the Mafia ran corporate America in the 1960s; more.
320 PAGES. 6X9 PAPERBACK. ILLUSTRATED. $16.00. CODE: IGF

KUNDALINI TALES
by Richard Sauder, Ph.D.

Underground Bases and Tunnels author Richard Sauder on his personal experiences and provocative research into spontaneous spiritual awakening, out-of-body journeys, encounters with secretive governmental powers, daylight sightings of UFOs, and more. Sauder continues his studies of underground bases with new information on the occult underpinnings of the U.S. space program. The book also contains a breakthrough section that examines actual U.S. patents for devices that manipulate minds and thoughts from a remote distance. Included are chapters on the secret space program and a 130-page appendix of patents and schematic diagrams of secret technology and mind control devices.
296 PAGES. 7X10 PAPERBACK. ILLUSTRATED. BIBLIOGRAPHY. $14.95. CODE: KTAL

LIQUID CONSPIRACY
JFK, LSD, the CIA, Area 51 & UFOs
by George Piccard

Underground author George Piccard on the politics of LSD, mind control, and Kennedy's involvement with Area 51 and UFOs. Reveals JFK's LSD experiences with Mary Pinchot-Meyer. The plot thickens with an ever expanding web of CIA involvement, from underground bases with UFOs seen by JFK and Marilyn Monroe (among others) to a vaster conspiracy that affects every government agency from NASA to the Justice Department. This may have been the reason that Marilyn Monroe and actress-columnist Dorothy Killgallen were both murdered. Focusing on the bizarre side of history, *Liquid Conspiracy* takes the reader on a psychedelic tour de force.
264 PAGES. 6X9 PAPERBACK. ILLUSTRATED. $14.95. CODE: LIQC

ATLANTIS: MOTHER OF EMPIRES
Atlantis Reprint Series
by Robert Stacy-Judd

Robert Stacy-Judd's classic 1939 book on Atlantis. Stacy-Judd was a California architect and an expert on the Mayas and their relationship to Atlantis. Stacy-Judd was an excellent artist and his book is lavishly illustrated. The eighteen comprehensive chapters in the book are: The Mayas and the Lost Atlantis; Conjectures and Opinions; The Atlantean Theory; Cro-Magnon Man; East Is West; And West Is East; The Mormons and the Mayas; Astrology in Two Hemispheres; The Language of Architecture; The American Indian; Pre-Panamanians and Pre-Incas; Columns and City Planning; Comparisons and Mayan Art; The Iberian Link; The Maya Tongue; Quetzalcoatl; Summing Up the Evidence; The Mayas in Yucatan.
340 PAGES. 8X11 PAPERBACK. ILLUSTRATED. INDEX. $19.95. CODE: AMOE

COSMIC MATRIX
Piece for a Jig-Saw, Part Two
by Leonard G. Cramp

Leonard G. Cramp, a British aerospace engineer, wrote his first book *Space Gravity and the Flying Saucer* in 1954. *Cosmic Matrix* is the long-awaited sequel to his 1966 book *UFOs & Anti-Gravity: Piece for a Jig-Saw*. Cramp has had a long history of examining UFO phenomena and has concluded that UFOs use the highest possible aeronautic science to move in the way they do. Cramp examines anti-gravity effects and theorizes that this super-science used by the craft—described in detail in the book—can lift mankind into a new level of technology, transportation and understanding of the universe. The book takes a close look at gravity control, time travel, and the interlocking web of energy between all planets in our solar system with Leonard's unique technical diagrams. A fantastic voyage into the present and future!
364 PAGES. 6X9 PAPERBACK. ILLUSTRATED. BIBLIOGRAPHY. $16.00. CODE: CMX

24 HOUR CREDIT CARD ORDERS—CALL: 815-253-6390 FAX: 815-253-6300
EMAIL: AUPHQ@FRONTIERNET.NET   HTTP://WWW.ADVENTURESUNLIMITED.CO.NZ

NEW BOOKS

THE TIME TRAVEL HANDBOOK
A Manual of Practical Teleportation & Time Travel
edited by David Hatcher Childress

In the tradition of *The Anti-Gravity Handbook* and *The Free-Energy Device Handbook*, science and UFO author David Hatcher Childress takes us into the weird world of time travel and teleportation. Not just a whacked-out look at science fiction, this book is an authoritative chronicling of real-life time travel experiments, teleportation devices and more. *The Time Travel Handbook* takes the reader beyond the government experiments and deep into the uncharted territory of early time travellers such as Nikola Tesla and Guglielmo Marconi and their alleged time travel experiments, as well as the Wilson Brothers of EMI and their connection to the Philadelphia Experiment—the U.S. Navy's forays into invisibility, time travel, and teleportation. Childress looks into the claims of time travelling individuals, and investigates the unusual claim that the pyramids on Mars were built in the future and sent back in time. A highly visual, large format book, with patents, photos and schematics. Be the first on your block to build your own time travel device!
316 PAGES. 7X10 PAPERBACK. ILLUSTRATED. $16.95. CODE: TTH. MAY PUBLICATION

PATH OF THE POLE
Cataclysmic Poles Shift Geology
by Charles Hapgood

Maps of the Ancient Sea Kings author Hapgood's classic book *Path of the Pole* is back in print! Hapgood researched Antarctica, ancient maps and the geological record to conclude that the Earth's crust has slipped in the inner core many times in the past, changing the position of the pole. *Path of the Pole* discusses the various "pole shifts" in Earth's past, giving evidence for each one, and moves on to possible future pole shifts. Packed with illustrations, this is the sourcebook for many other books on cataclysms and pole shifts such as *5-5-2000: Ice the Ultimate Disaster* by Richard Noone. A planetary alignment on May 5, 2000 is predicted to cause the next pole shift—a date that is less than a year away! With Millennium Madness in full swing, this is sure to be a popular book.
356 PAGES. 6X9 PAPERBACK. ILLUSTRATED. $16.95. CODE: POP. MAY PUBLICATION

IN SEARCH OF ADVENTURE
A Wild Travel Anthology
compiled by Bruce Northam & Brad Olsen

An epic collection of 100 travelers' tales—a compendium that celebrates the wild side of contemporary travel writing—relating humorous, revealing, sometimes naughty stories by acclaimed authors. Indeed, a book to heat up the gypsy blood in all of us. Stories by Tim Cahill, Simon Winchester, Marybeth Bond, Robert Young Pelton, David Hatcher Childress, Richard Bangs, Linda Watanabe McFerrin, Jorma Kaukonen, and many more.
459 PAGES. 6X9 PAPERBACK. ILLUSTRATED. $17.95. CODE: ISOA

ECCENTRIC LIVES AND PECULIAR NOTIONS
by John Michell

The first paperback edition of Michell's fascinating study of the lives and beliefs of over 20 eccentric people. Published in hardback by Thames & Hudson in London, *Eccentric Lives and Peculiar Notions* takes us into the bizarre and often humorous lives of such people as Lady Blount, who was sure that the earth is flat; Cyrus Teed, who believed that the earth is a hollow shell with us on the inside; Edward Hine, who believed that the British are the lost Tribes of Israel; and Baron de Guldenstubbe, who was sure that statues wrote him letters. British writer and housewife Nesta Webster devoted her life to exposing international conspiracies, and Father O'Callaghan devoted his to opposing interest on loans. The extraordinary characters in this book were—and in some cases still are—wholehearted enthusiasts for the various causes and outrageous notions they adopted, and John Michell describes their adventures with spirit and compassion. Some of them prospered and lived happily with their obsessions, while others failed dismally. We read of the hapless inventor of a giant battleship made of ice who died alone and neglected, and of the London couple who achieved peace and prosperity by drilling holes in their heads. Other chapters on the Last of the Welsh Druids; Congressman Ignacius Donnelly, the Great Heretic and Atlantis; Shakespearean Decoders and the Baconian Treasure Hunt; Early Ufologists; Jerusalem in Scotland; Bibliomaniacs; more.
248 PAGES. 6X9 PAPERBACK. ILLUSTRATED. $14.95. CODE: ELPN. MAY PUBLICATION

THE CHRIST CONSPIRACY
The Greatest Story Ever Sold
by Acharya S.

In this highly controversial and explosive book, archaeologist, historian, mythologist and linguist Acharya S. marshals an enormous amount of startling evidence to demonstrate that Christianity and the story of Jesus Christ were created by members of various secret societies, mystery schools and religions in order to unify the Roman Empire under one state religion. In developing such a fabrication, this multinational cabal drew upon a multitude of myths and rituals that existed long before the Christian era, and reworked them for centuries into the religion passed down to us today. Contrary to popular belief, there was no single man who was at the genesis of Christianity; Jesus was many characters rolled into one. These characters personified the ubiquitous solar myth, and their exploits were well known, as reflected by such popular deities as Mithras, Heracles/Hercules, Dionysos and many others throughout the Roman Empire and beyond. The story of Jesus as portrayed in the Gospels is revealed to be nearly identical in detail to that of the earlier savior-gods Krishna and Horus, who for millennia preceding Christianity held great favor with the people. *The Christ Conspiracy* shows the Jesus character as not unique or original, not "divine revelation." Christianity re-interprets the same extremely ancient body of knowledge that revolved around the celestial bodies and natural forces. The result of this myth making has been "The Greatest Conspiracy Ever Sold."
256 PAGES. 6X9 PAPERBACK. ILLUSTRATED. $14.95. CODE: CHRC. JUNE PUBLICATION

24 HOUR CREDIT CARD ORDERS—CALL: 815-253-6390 FAX: 815-253-6300
EMAIL: AUPHQ@FRONTIERNET.NET   HTTP://WWW.ADVENTURESUNLIMITED.CO.NZ

ANTI-GRAVITY

THE ANTI-GRAVITY HANDBOOK
edited by David Hatcher Childress, with Nikola Tesla, T.B. Paulicki, Bruce Cathie, Albert Einstein and others

The new expanded compilation of material on Anti-Gravity, Free Energy, Flying Saucer Propulsion, UFOs, Suppressed Technology, NASA Cover-ups and more. Highly illustrated with patents, technical illustrations and photos. This revised and expanded edition has more material, including photos of Area 51, Nevada, the government's secret testing facility. This classic on weird science is back in a 90s format!
- How to build a flying saucer.
- Crystals and their role in levitation.
- Secret government research and development.
- Nikola Tesla on how anti-gravity airships could draw power from the atmosphere.
- Bruce Cathie's Anti-Gravity Equation.
- NASA, the Moon and Anti-Gravity.

230 PAGES. 7X10 PAPERBACK. BIBLIOGRAPHY. APPENDIX. ILLUSTRATED. $14.95. CODE: AGH

ANTI-GRAVITY & THE WORLD GRID
edited by David Hatcher Childress

Is the earth surrounded by an intricate electromagnetic grid network offering free energy? This compilation of material on ley lines and world power points contains chapters on the geography, mathematics, and light harmonics of the earth grid. Learn the purpose of ley lines and ancient megalithic structures located on the grid. Discover how the grid made the Philadelphia Experiment possible. Explore the Coral Castle and many other mysteries, including acoustic levitation, Tesla Shields and scalar wave weaponry. Browse through the section on anti-gravity patents, and research resources.

274 PAGES. 7X10 PAPERBACK. ILLUSTRATED. $14.95. CODE: AGW

ANTI-GRAVITY & THE UNIFIED FIELD
edited by David Hatcher Childress

Is Einstein's Unified Field Theory the answer to all of our energy problems? Explored in this compilation of material is how gravity, electricity and magnetism manifest from a unified field around us. Why artificial gravity is possible; secrets of UFO propulsion; free energy; Nikola Tesla and anti-gravity airships of the 20s and 30s; flying saucers as superconducting whirls of plasma; anti-mass generators; vortex propulsion; suppressed technology; government cover-ups; gravitational pulse drive; spacecraft & more.

240 PAGES. 7X10 PAPERBACK. ILLUSTRATED. $14.95. CODE: AGU

ETHER TECHNOLOGY
A Rational Approach to Gravity Control
by Rho Sigma

This classic book on anti-gravity and free energy is back in print. Written by a well-known American scientist under the pseudonym of "Rho Sigma," this book delves into international efforts at gravity control and discoid craft propulsion. Before the Quantum Field, there was "Ether." This small, but informative book has chapters on John Searle and "Searle discs;" T. Townsend Brown and his work on anti-gravity and ether-vortex turbines. Includes a forward by former NASA astronaut Edgar Mitchell.

108 PAGES. 6X9 PAPERBACK. ILLUSTRATED. $12.95. CODE: ETT

MAN-MADE UFOS 1944—1994
Fifty Years of Suppression
by Renato Vesco & David Hatcher Childress

A comprehensive look at the early "flying saucer" technology of Nazi Germany and the genesis of man-made UFOs. This book takes us from the work of captured German scientists to escaped battalions of Germans, secret communities in South America and Antarctica to today's state-of-the-art "Dreamland" flying machines. Heavily illustrated, this astonishing book blows the lid off the "government UFO conspiracy" and explains with technical diagrams the technology involved. Examined in detail are secret underground airfields and factories; German secret weapons; "suction" aircraft; the origin of NASA; gyroscopic stabilizers and engines; the secret Marconi aircraft factory in South America; and more. Not to be missed by students of technology suppression, secret societies, anti-gravity, free energy, conspiracy and World War II! Introduction by W.A. Harbinson, author of the Dell novels *GENESIS* and *REVELATION*.

318 PAGES. 6X9 PAPERBACK. ILLUSTRATED. INDEX & FOOTNOTES. $18.95. CODE: MMU

---

24 HOUR CREDIT CARD ORDERS—CALL: 815-253-6390 FAX: 815-253-6300
EMAIL: AUPHQ@FRONTIERNET.NET HTTP://WWW.ADVENTURESUNLIMITED.CO.NZ

FREE ENERGY SYSTEMS

THE FREE-ENERGY DEVICE HANDBOOK
A Compilation of Patents and Reports
by David Hatcher Childress
A large-format compilation of various patents, papers, descriptions and diagrams concerning free energy devices and systems. *The Free-Energy Device Handbook* is a visual tool for experimenters and researchers into magnetic motors and other "over-unity" devices. With chapters on the Adams Motor, the Hans Coler Generator, cold fusion, superconductors, "N" machines, space-energy generators, Nikola Tesla, T. Townsend Brown, and the latest in free energy devices. Packed with photos, technical diagrams, patents and fascinating information, this book belongs on every science shelf. With energy and profit being major political reasons for fighting wars, free energy devices, if ever allowed to be mass distributed to consumers, could change the world! Get your copy now before the Department of Energy bans this book!
292 PAGES. 8X10 PAPERBACK. ILLUSTRATED. BIBLIOGRAPHY. $16.95. CODE: FEH

UFOS AND ANTI-GRAVITY
Piece for a Jig-Saw
by Leonard G. Cramp
Leonard G. Cramp's 1966 classic book on flying saucer propulsion and suppressed technology is available again. *UFOS & Anti-Gravity: Piece for a Jig-Saw* is a highly technical look at the UFO phenomenon by a trained scientist. Cramp first introduces the idea of 'anti-gravity' and introduces us to the various theories of gravitation. He then examines the technology necessary to build a flying saucer and examines in great detail the technical aspects of such a craft. Cramp's book is a wealth of material and diagrams on flying saucers, anti-gravity, suppressed technology, G-fields and UFOs. Chapters include Crossroads of Aerodymanics, Aerodynamic Saucers, Limitations of Rocketry, Gravitation and the Ether, Gravitational Spaceships, G-Field Lift Effects, The Bi-Field Theory, VTOL and Hovercraft, Analysis of UFO photos, more. "I feel the Air Force has not been giving out all available information on these unidentified flying objects. You cannot disregard so many unimpeachable sources." — John McCormack, Speaker of the U.S. House of Representatives.
388 PAGES. 6X9 PAPERBACK. HEAVILY ILLUSTRATED. $16.95. CODE: UAG

THE HARMONIC CONQUEST OF SPACE
by Captain Bruce Cathie
A new, updated edition with additional material. Chapters include: Mathematics of the World Grid; the Harmonics of Hiroshima and Nagasaki; Harmonic Transmission and Receiving; the Link Between Human Brain Waves; the Cavity Resonance between the Earth; the Ionosphere and Gravity; Edgar Cayce—the Harmonics of the Subconscious; Stonehenge, the Harmonics of the Moon; the Pyramids of Mars; Nikola Tesla's Electric Car; the Robert Adams Pulsed Electric Motor Generator; Harmonic Clues to the Unified Field; and more. Also included are tables showing the harmonic relations between the earth's magnetic field, the speed of light, and anti-gravity/gravity acceleration at different points on the earth's surface. New chapters in this edition on the giant stone spheres of Costa Rica, Atomic Tests and Volcanic Activity, and a chapter on Ayers Rock analyzed with Stone Mountain, Georgia.
248 PAGES. 6X9. PAPERBACK. ILLUSTRATED. BIBLIOGRAPHY. $16.95. CODE: HCS

THE ENERGY GRID
Harmonic 695, The Pulse of the Universe
by Captain Bruce Cathie.
This is the breakthrough book that explores the incredible potential of the Energy Grid and the Earth's Unified Field all around us. Cathie's first book, *Harmonic 33*, was published in 1968 when he was a commercial pilot in New Zealand. Since then, Captain Bruce Cathie has been the premier investigator into the amazing potential of the infinite energy that surrounds our planet every microsecond. Cathie investigates the Harmonics of Light and how the Energy Grid is created. In this amazing book are chapters on UFO Propulsion, Nikola Tesla, Unified Equations, the Mysterious Aerials, Pythagoras & the Grid, Nuclear Detonation and the Grid, Maps of the Ancients, an Australian Stonehenge examined, more.
255 PAGES. 6X9 PAPERBACK. ILLUSTRATED. $15.95. CODE: TEG

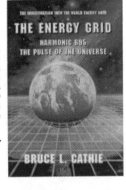

THE BRIDGE TO INFINITY
Harmonic 371244
by Captain Bruce Cathie
Cathie has popularized the concept that the earth is crisscrossed by an electromagnetic grid system that can be used for anti-gravity, free energy, levitation and more. This book includes a new analysis of the harmonic nature of reality, acoustic levitation, pyramid power, harmonic receiver towers and UFO propulsion. It concludes that today's scientists have at their command a fantastic store of knowledge with which to advance the welfare of the human race. Chapters on The Harmonics of the Philadelphia Experiment; of Tunguska; of Krakatoa; the Pyramids of Shensi; of Stonehenge; Adams Free Energy Motor; more.
204 PAGES. 6X9 PAPERBACK. ILLUSTRATED. $14.95. CODE: BTF

24 HOUR CREDIT CARD ORDERS—CALL: 815-253-6390 FAX: 815-253-6300
EMAIL: AUPHQ@FRONTIERNET.NET HTTP://WWW.ADVENTURESUNLIMITED.CO.NZ

ANTI-GRAVITY

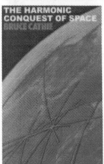

THE HARMONIC CONQUEST OF SPACE
NEW EDITION!
by Captain Bruce Cathie
A new, updated edition with additional material. Chapters include: Mathematics of the World Grid; the Harmonics of Hiroshima and Nagasaki; Harmonic Transmission and Receiving; the Link Between Human Brain Waves; the Cavity Resonance between the Earth; the Ionosphere and Gravity; Edgar Cayce—the Harmonics of the Subconscious; Stonehenge; the Harmonics of the Moon; the Pyramids of Mars; Nikola Tesla's Electric Car; the Robert Adams Pulsed Electric Motor Generator; Harmonic Clues to the Unified Field; and more. Also included are tables showing the harmonic relations between the earth's magnetic field, the speed of light, and anti-gravity/gravity acceleration at different points on the earth's surface. New chapters in this edition on the giant stone spheres of Costa Rica, Atomic Tests and Volcanic Activity, and a chapter on Ayers Rock analysed with Stone Mountain, Georgia.
248 PAGES. 6X9. PAPERBACK. ILLUSTRATED. BIBLIOGRAPHY. $16.95. CODE: HCS

THE BRIDGE TO INFINITY
Harmonic 371244
by Captain Bruce Cathie
Cathie has popularized the concept that the earth is crisscrossed by an electromagnetic grid system that can be used for anti-gravity, free energy, levitation and more. The book includes a new analysis of the harmonic nature of reality, acoustic levitation, pyramid power, harmonic receiver towers and UFO propulsion. It concludes that today's scientists have at their command a fantastic store of knowledge with which to advance the welfare of the human race.
204 PAGES. 6X9 TRADEPAPER. ILLUSTRATED. $14.95. CODE: BTF

THE ENERGY GRID
Harmonic 695, The Pulse of the Universe
by Captain Bruce Cathie.
This is the breakthrough book that explores the incredible potential of the Energy Grid and the Earth's Unified Field all around us. Cathie's first book, *Harmonic 33*, was published in 1968 when he was a commercial pilot in New Zealand. Since then, Captain Bruce Cathie has been the premier investigator into the amazing potential of the infinite energy that surrounds our planet every microsecond. Cathie investigates the Harmonics of Light and how the Energy Grid is created. In this amazing book are chapters on UFO Propulsion, Nikola Tesla, Unified Equations, the Mysterious Aerials, Pythagoras & the Grid, Nuclear Detonation and the Grid, Maps of the Ancients, an Australian Stonehenge examined, more.
255 PAGES. 6X9 TRADEPAPER. ILLUSTRATED. $15.95. CODE: TEG

THE GIFT
The Crop Circles Deciphered
by Doug Ruby
A fascinating and well-illustrated book on the author's attempt to decipher the crop circles of England and his conclusion that they represent some sort of spinning magnet motor! Ruby is an airline pilot with an interest in these things and this book is definitely a must for everyone with interest in free-energy motors, anti-gravity, crop-circles, the UFO enigma and more. A deluxe book with many of the illustrations in color!
174 PAGES. 7X9 HARDBACK. ILLUSTRATED. REFERENCES. $32.95. CODE: GIFT

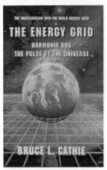

FIELD EFFECT
...The Pi Phase of Physics & The Unified Field
by Leigh Richmond-Donahue
This book is an analysis of the structure of the electron which brings to light the anatomy of black holes, superstrings, and the techniques of evolution. Leigh shows how the "field" is complete around the universe and all living things. Our galaxy, ourselves, and the electrons of which we are built, function on a logical pattern and operative system. Once you find that pattern, the rest falls into place, including free energy, anti-gravity and human enlightenment.
77 PAGES. 6X9 TRADEPAPER. ILLUSTRATED. $11.95. CODE: FEF

THE VORTEX
Key to Future Science
by David Ash & Peter Hewitt
The authors' rediscovery of the forgotten vision of an outstanding Scottish scientist promises a revolution which could change our thinking about everything. Pointing to a bridge between the physical and non-physical worlds, it opens the frontiers of science to the supernatural in a way never before possible. *The Vortex* also demonstrates the immense powers available through nature to those working on subtle energy levels. Chapters on Lord Kelvin's Vortex, UFOs Explained, The Return of Pan, The Dimension of the Gods, more.
192 PAGES. 6X9 PAPERBACK. ILLUSTRATED. $14.95. CODE: VOR

24 HOUR CREDIT CARD ORDERS—CALL: 815-253-6390 FAX: 815-253-6300
EMAIL: AUPHQ@FRONTIERNET.NET HTTP://WWW.ADVENTURESUNLIMITED.CO.NZ

CONSPIRACY & HISTORY

HAARP
The Ultimate Weapon of the Conspiracy
by Jerry Smith

The HAARP project in Alaska is one of the most controversial projects ever undertaken by the U.S. Government. Jerry Smith gives us the history of the HAARP project and explains how it can be used as an awesome weapon of destruction. Smith exposes a covert military project and the web of conspiracies behind it. HAARP has many possible scientific and military applications, from raising a planetary defense shield to peering deep into the earth. Smith leads the reader down a trail of solid evidence into ever deeper and scarier conspiracy theories in an attempt to discover the "whos" and "whys" behind HAARP, and discloses a possible plan to rule the world. At best, HAARP is science out-of-control; at worst, HAARP could be the most dangerous device ever created, a futuristic technology that is everything from super-beam weapon to world-wide mind control device. The Star Wars future is now! Topics include Over-the-Horizon Radar and HAARP, Mind Control, ELF and HAARP, The Telsa Connection, The Russian Woodpecker, GWEN & HAARP, Earth Penetrating Tomography, Weather Modification, Secret Science of the Conspiracy, more. Includes the complete 1987 Bernard Eastlund patent for his pulsed super-weapon that he claims was stolen by the HAARP Project.
256 PAGES. 6X9 PAPERBACK. ILLUSTRATED. $14.95. CODE: HARP

MIND CONTROL, OSWALD & JFK:
Were We Controlled?
introduction by Kenn Thomas

Steamshovel Press editor Kenn Thomas examines the little-known book *Were We Controlled?*, first published in 1968. The book maintained that Lee Harvey Oswald was a special agent who was a mind control subject, having received an implant in 1960 at a Russian hospital. Thomas examines the evidence for implant technology and the role it could have played in the Kennedy Assassination. Thomas also looks at the mind control aspects of the RFK assassination and details the history of implant technology. A growing number of people are interested in CIA experiments and its "Silent Weapons for Quiet Wars."
256 PAGES. 6X9 PAPERBACK. ILLUSTRATED. NOTES. $16.00. CODE: MCOJ

MIND CONTROL, WORLD CONTROL
by Jim Keith

Veteran author and investigator Jim Keith uncovers a surprising amount of information on the technology, experimentation and implementation of mind control. Various chapters in this shocking book are on early CIA experiments such as Project Artichoke and Project R.H.I.C.-EDOM, the methodology and technology of implants, mind control assassins and couriers, various famous Mind Control victims such as Sirhan Sirhan and Candy Jones. Also featured in this book are chapters on how mind control technology may be linked to some UFO activity and "UFO abductions."
256 PAGES. 6X9 PAPERBACK. ILLUSTRATED. FOOTNOTES. $14.95. CODE: MCWC

PROJECT SEEK
Onassis, Kennedy and the Gemstone Thesis
by Gerald A. Carroll

This book reprints the famous Gemstone File, a document circulated in 1974 concerning the Mafia, Onassis and the Kennedy assassination. With the passing of Jackie Kennedy-Onassis, this information on the Mafia and the CIA, the formerly "Hughes" controlled defense industry, and the violent string of assassinations can at last be told. Also includes new information on the Nugan Hand Bank, the BCCI scandal, "the Octopus," and the Paul Wilcher Transcripts.
388 PAGES. 6X9 PAPERBACK. ILLUSTRATED. $16.95. CODE: PJS

NASA, NAZIS & JFK:
The Torbitt Document & the JFK Assassination
introduction by Kenn Thomas

This book emphasizes the links between "Operation Paper Clip" Nazi scientists working for NASA, the assassination of JFK, and the secret Nevada air base Area 51. The Torbitt Document also talks about the roles played in the assassination by Division Five of the FBI, the Defense Industrial Security Command (DISC), the Las Vegas mob, and the shadow corporate entities Permindex and Centro-Mondiale Commerciale. The Torbitt Document claims that the same players planned the 1962 assassination attempt on Charles de Gaul, who ultimately pulled out of NATO because he traced the "Assassination Cabal" to Permindex in Switzerland and to NATO headquarters in Brussels. The Torbitt Document paints a dark picture of NASA, the military industrial complex, and the connections to Mercury, Nevada which headquarters the "secret space program."
258 PAGES. 5X8. PAPERBACK. ILLUSTRATED. $16.00. CODE: NNJ

THE HISTORY OF THE KNIGHTS TEMPLARS
The Temple Church and the Temple
by Charles G. Addison, introduction by David Hatcher Childress

Chapters on the origin of the Templars, their popularity in Europe and their rivalry with the Knights of St. John, later to be known as the Knights of Malta. Detailed information on the activities of the Templars in the Holy Land, and the 1312 AD suppression of the Templars in France and other countries, which culminated in the execution of Jacques de Molay and the continuation of the Knights Templars in England and Scotland; the formation of the society of Knights Templars in London; and the rebuilding of the Temple in 1816. Plus a lengthy intro about the lost Templar fleet and its connections to the ancient North American sea routes.
395 PAGES. 6X9 PAPERBACK. ILLUSTRATED. $16.95. CODE: HKT

---

24 HOUR CREDIT CARD ORDERS—CALL: 815-253-6390 FAX: 815-253-6300
EMAIL: AUPHQ@FRONTIERNET.NET HTTP://WWW.ADVENTURESUNLIMITED.CO.NZ

THE LOST CITIES SERIES

VIMANA AIRCRAFT OF ANCIENT INDIA & ATLANTIS
by David Hatcher Childress
introduction by Ivan T. Sanderson

Did the ancients have the technology of flight? In this incredible volume on ancient India, authentic Indian texts such as the *Ramayana* and the *Mahabharata* are used to prove that ancient aircraft were in use more than four thousand years ago. Included in this book is the entire Fourth Century BC manuscript *Vimaanika Shastra* by the ancient author Maharishi Bharadwaaja, translated into English by the Mysore Sanskrit professor G.R. Josyer. Also included are chapters on Atlantean technology, the incredible Rama Empire of India and the devastating wars that destroyed it. Also an entire chapter on mercury vortex propulsion and mercury gyros, the power source described in the ancient Indian texts. Not to be missed by those interested in ancient civilizations or the UFO enigma.

334 PAGES. 6X9 PAPERBACK. RARE PHOTOGRAPHS, MAPS AND DRAWINGS. $15.95. CODE: VAA

LOST CONTINENTS & THE HOLLOW EARTH
I Remember Lemuria and the Shaver Mystery
by David Hatcher Childress & Richard Shaver

Lost Continents & the Hollow Earth is Childress' thorough examination of the early hollow earth stories of Richard Shaver and the fascination that fringe fantasy subjects such as lost continents and the hollow earth have had for the American public. Shaver's rare 1948 book *I Remember Lemuria* is reprinted in its entirety, and the book is packed with illustrations from Ray Palmer's *Amazing Stories* magazine of the 1940s. Palmer and Shaver told of tunnels running through the earth—tunnels inhabited by the Deros and Teros, humanoids from an ancient spacefaring race that had inhabited the earth, eventually going underground, hundreds of thousands of years ago. Childress discusses the famous hollow earth books and delves deep into whatever reality may be behind the stories of tunnels in the earth. Operation High Jump to Antarctica in 1947 and Admiral Byrd's bizarre statements, tunnel systems in South America and Tibet, the underground world of Agartha, the belief of UFOs coming from the South Pole, more.

344 PAGES. 6X9 PAPERBACK. ILLUSTRATED. $16.95. CODE: LCHE

LOST CITIES OF NORTH & CENTRAL AMERICA
by David Hatcher Childress

Down the back roads from coast to coast, maverick archaeologist and adventurer David Hatcher Childress goes deep into unknown America. With this incredible book, you will search for lost Mayan cities and books of gold, discover an ancient canal system in Arizona, climb gigantic pyramids in the Midwest, explore megalithic monuments in New England, and join the astonishing quest for lost cities throughout North America. From the war-torn jungles of Guatemala, Nicaragua and Honduras to the deserts, mountains and fields of Mexico, Canada, and the U.S.A., Childress takes the reader in search of sunken ruins, Viking forts, strange tunnel systems, living dinosaurs, early Chinese explorers, and fantastic lost treasure. Packed with both early and current maps, photos and illustrations.

590 PAGES. 6X9 PAPERBACK. PHOTOS, MAPS, AND ILLUSTRATIONS. FOOTNOTES & BIBLIOGRAPHY. $14.95. CODE: NCA

LOST CITIES & ANCIENT MYSTERIES OF SOUTH AMERICA
by David Hatcher Childress

Rogue adventurer and maverick archaeologist David Hatcher Childress takes the reader on unforgettable journeys deep into deadly jungles, high up on windswept mountains and across scorching deserts in search of lost civilizations and ancient mysteries. Travel with David and explore stone cities high in mountain forests and hear fantastic tales of Inca treasure, living dinosaurs, and a mysterious tunnel system. Whether he is hopping freight trains, searching for secret cities, or just dealing with the daily problems of food, money, and romance, the author keeps the reader spellbound. Includes both early and current maps, photos, and illustrations, and plenty of advice for the explorer planning his or her own journey of discovery.

381 PAGES. 6X9 PAPERBACK. PHOTOS, MAPS, AND ILLUSTRATIONS. FOOTNOTES & BIBLIOGRAPHY. $14.95. CODE: SAM

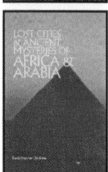

LOST CITIES & ANCIENT MYSTERIES OF AFRICA & ARABIA
by David Hatcher Childress

Across ancient deserts, dusty plains and steaming jungles, maverick archaeologist David Childress continues his world-wide quest for lost cities and ancient mysteries. Join him as he discovers forbidden cities in the Empty Quarter of Arabia; "Atlantean" ruins in Egypt and the Kalahari desert; a mysterious, ancient empire in the Sahara; and more. This is the tale of an extraordinary life on the road: across war-torn countries, Childress searches for King Solomon's Mines, living dinosaurs, the Ark of the Covenant and the solutions to some of the fantastic mysteries of the past.

423 PAGES. 6X9 PAPERBACK. PHOTOS, MAPS, AND ILLUSTRATIONS. FOOTNOTES & BIBLIOGRAPHY. $14.95. CODE: AFA

---

24 HOUR CREDIT CARD ORDERS—CALL: 815-253-6390 FAX: 815-253-6300
EMAIL: AUPHQ@FRONTIERNET.NET HTTP://WWW.ADVENTURESUNLIMITED.CO.NZ

THE LOST CITIES SERIES

LOST CITIES OF ATLANTIS, ANCIENT EUROPE & THE MEDITERRANEAN
by David Hatcher Childress
Atlantis! The legendary lost continent comes under the close scrutiny of maverick archaeologist David Hatcher Childress in this sixth book in the internationally popular *Lost Cities* series. Childress takes the reader in search of sunken cities in the Mediterranean; across the Atlas Mountains in search of Atlantean ruins; to remote islands in search of megalithic ruins; to meet living legends and secret societies. From Ireland to Turkey, Morocco to Eastern Europe, and around the remote islands of the Mediterranean and Atlantic, Childress takes the reader on an astonishing quest for mankind's past. Ancient technology, cataclysms, megalithic construction, lost civilizations and devastating wars of the past are all explored in this book. Childress challenges the skeptics and proves that great civilizations not only existed in the past, but the modern world and its problems are reflections of the ancient world of Atlantis.
524 PAGES. 6X9 PAPERBACK. ILLUSTRATED WITH 100S OF MAPS, PHOTOS AND DIAGRAMS. BIBLIOGRAPHY & INDEX. $16.95. CODE: MED

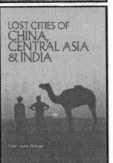

LOST CITIES OF CHINA, CENTRAL INDIA & ASIA
by David Hatcher Childress
Like a real life "Indiana Jones," maverick archaeologist David Childress takes the reader on an incredible adventure across some of the world's oldest and most remote countries in search of lost cities and ancient mysteries. Discover ancient cities in the Gobi Desert; hear fantastic tales of lost continents, vanished civilizations and secret societies bent on ruling the world; visit forgotten monasteries in forbidding snow-capped mountains with strange tunnels to mysterious subterranean cities! A unique combination of far-out exploration and practical travel advice, it will astound and delight the experienced traveler or the armchair voyager.
429 PAGES. 6X9 PAPERBACK. ILLUSTRATED. FOOTNOTES & BIBLIOGRAPHY. $14.95. CODE: CHI

LOST CITIES OF ANCIENT LEMURIA & THE PACIFIC
by David Hatcher Childress
Was there once a continent in the Pacific? Called Lemuria or Pacifica by geologists, Mu or Pan by the mystics, there is now ample mythological, geological and archaeological evidence to "prove" that an advanced and ancient civilization once lived in the central Pacific. Maverick archaeologist and explorer David Hatcher Childress combs the Indian Ocean, Australia and the Pacific in search of the surprising truth about mankind's past. Contains photos of the underwater city on Pohnpei; explanations on how the statues were levitated around Easter Island in a clockwise vortex movement; tales of disappearing islands; Egyptians in Australia; and more.
379 PAGES. 6X9 PAPERBACK. ILLUSTRATED. FOOTNOTES & BIBLIOGRAPHY. $14.95. CODE: LEM

ANCIENT TONGA
& the Lost City of Mu'a
by David Hatcher Childress
Lost Cities series author Childress takes us to the south sea islands of Tonga, Rarotonga, Samoa and Fiji to investigate the megalithic ruins on these beautiful islands. The great empire of the Polynesians, centered on Tonga and the ancient city of Mu'a, is revealed with old photos, drawings and maps. Chapters in this book are on the Lost City of Mu'a and its many megalithic pyramids, the Ha'amonga Trilithon and ancient Polynesian astronomy, Samoa and the search for the lost land of Havai'iki, Fiji and its wars with Tonga, Rarotonga's megalithic road, and Polynesian cosmology. Material on Egyptians in the Pacific, earth changes, the fortified moat around Mu'a, lost roads, more.
218 PAGES. 6X9 PAPERBACK. ILLUSTRATED. COLOR PHOTOS. BIBLIOGRAPHY. $15.95. CODE: TONG

ANCIENT MICRONESIA
& the Lost City of Nan Madol
by David Hatcher Childress

Micronesia, a vast archipelago of islands west of Hawaii and south of Japan, contains some of the most amazing megalithic ruins in the world. Part of our *Lost Cities* series, this volume explores the incredible conformations on various Micronesian islands, especially the fantastic and little-known ruins of Nan Madol on Pohnpei Island. The huge canal city of Nan Madol contains over 250 million tons of basalt columns over an 11 square-mile area of artificial islands. Much of the huge city is submerged, and underwater structures can be found to an estimated 80 feet. Islanders' legends claim that the basalt rocks, weighing up to 50 tons, were magically levitated into place by the powerful forefathers. Other ruins in Micronesia that are profiled include the Latte Stones of the Marianas, the menhirs of Palau, the megalithic canal city on Kosrae Island, megaliths on Guam, and more.
256 PAGES. 6X9 PAPERBACK. ILLUSTRATED. INCLUDES A COLOR PHOTO SECTION. BIBLIOGRAPHY. $16.95. CODE: AMIC

---

24 HOUR CREDIT CARD ORDERS—CALL: 815-253-6390 FAX: 815-253-6300
EMAIL: AUPHQ@FRONTIERNET.NET HTTP://WWW.ADVENTURESUNLIMITED.CO.NZ

ATLANTIS REPRINT SERIES

ATLANTIS: MOTHER OF EMPIRES
Atlantis Reprint Series
by Robert Stacy-Judd
Robert Stacy-Judd's classic 1939 book on Atlantis is back in print in this large-format paperback edition. Stacy-Judd was a California architect and an expert on the Mayas and their relationship to Atlantis. He was an excellent artist and his work is lavishly illustrated. The eighteen comprehensive chapters in the book are: The Mayas and the Lost Atlantis; Conjectures and Opinions; The Atlantean Theory; Cro-Magnon Man; East is West; And West is East; The Mormons and the Mayas; Astrology in Two Hemispheres; The Language of Architecture; The American Indian; Pre-Panamanians and Pre-Incas; Columns and City Planning; Comparisons and Mayan Art; The Iberian Link; The Maya Tongue; Quetzalcoatl; Summing Up the Evidence; The Mayas in Yucatan.
340 PAGES. 8X11 PAPERBACK. ILLUSTRATED. INDEX. $19.95. CODE: AMOE

SECRET CITIES OF OLD SOUTH AMERICA
Atlantis Reprint Series
by Harold T. Wilkins
The reprint of Wilkins' classic book, first published in 1952, claiming that South America was Atlantis. Chapters include Mysteries of a Lost World; Atlantis Unveiled; Red Riddles on the Rocks; South America's Amazons Existed!; The Mystery of El Dorado and Gran Payatiti—the Final Refuge of the Incas; Monstrous Beasts of the Unexplored Swamps & Wilds; Weird Denizens of Antediluvian Forests; New Light on Atlantis from the World's Oldest Book; The Mystery of Old Man Noah and the Arks; and more.
438 PAGES. 6X9 PAPERBACK. ILLUSTRATED. BIBLIOGRAPHY & INDEX. $16.95. CODE: SCOS

THE SHADOW OF ATLANTIS
The Echoes of Atlantean Civilization Tracked through Space & Time
by Colonel Alexander Braghine
First published in 1940, *The Shadow of Atlantis* is one of the great classics of Atlantis research. The book amasses a great deal of archaeological, anthropological, historical and scientific evidence in support of a lost continent in the Atlantic Ocean. Braghine covers such diverse topics as Egyptians in Central America, the myth of Quetzalcoatl, the Basque language and its connection with Atlantis, the connections with the ancient pyramids of Mexico, Egypt and Atlantis, the sudden demise of mammoths, legends of giants and much more. Braghine was a linguist and spends part of the book tracing ancient languages to Atlantis and studying little-known inscriptions in Brazil, deluge myths and the connections between ancient languages. Braghine takes us on a fascinating journey through space and time in search of the lost continent.
288 PAGES. 6X9 PAPERBACK. ILLUSTRATED. $16.95. CODE: SOA

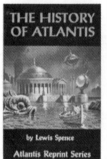

RIDDLE OF THE PACIFIC
by John Macmillan Brown
Oxford scholar Brown's classic work on lost civilizations of the Pacific is now back in print! John Macmillan Brown was an historian and New Zealand's premier scientist when he wrote about the origins of the Maoris. After many years of travel thoughout the Pacific studying the people and customs of the south seas islands, he wrote *Riddle of the Pacific* in 1924. The book is packed with rare turn-of-the-century illustrations. Don't miss Brown's classic study of Easter Island, ancient scripts, megalithic roads and cities, more. Brown was an early believer in a lost continent in the Pacific.
460 PAGES. 6X9 PAPERBACK. ILLUSTRATED. $16.95. CODE: ROP

THE HISTORY OF ATLANTIS
by Lewis Spence
Lewis Spence's classic book on Atlantis is now back in print! Spence was a Scottish historian (1874-1955) who is best known for his volumes on world mythology and his five Atlantis books. *The History of Atlantis* (1926) is considered his finest. Spence does his scholarly best in chapters on the Sources of Atlantean History, the Geography of Atlantis, the Races of Atlantis, the Kings of Atlantis, the Religion of Atlantis, the Colonies of Atlantis, more. Sixteen chapters in all.
240 PAGES. 6X9 PAPERBACK. ILLUSTRATED WITH MAPS, PHOTOS & DIAGRAMS. $16.95. CODE: HOA

ATLANTIS IN SPAIN
A Study of the Ancient Sun Kingdoms of Spain
by E.M. Whishaw
First published by Rider & Co. of London in 1928, this classic book is a study of the megaliths of Spain, ancient writing, cyclopean walls, sun worshipping empires, hydraulic engineering, and sunken cities. An extremely rare book, it was out of print for 60 years. Learn about the Biblical Tartessus; an Atlantean city at Niebla; the Temple of Hercules and the Sun Temple of Seville; Libyans and the Copper Age; more. Profusely illustrated with photos, maps and drawings.
284 PAGES. 6X9 PAPERBACK. ILLUSTRATED. TABLES OF ANCIENT SCRIPTS. $15.95. CODE: AIS

24 HOUR CREDIT CARD ORDERS—CALL: 815-253-6390 FAX: 815-253-6300
EMAIL: AUPHQ@FRONTIERNET.NET HTTP://WWW.ADVENTURESUNLIMITED.CO.NZ

WHAT IS YOUR IDEA OF EXCITEMENT?   **OUR LATEST ISSUE**

√ **DIVING** to explore mysterious underwater ruins?
√ **HACKING** through thick jungle in searh of lost cities?
√ **TREKKING** shifting sands to a lost city in the Kalahari?
√ **CLIMBING** the Himalayas to remote monasteries?
√ **DIGGING** for treasure on remote tropical islands?
√ **DISCOVERING** previously unknown animal species?

AS A MEMBER OF THE WORLD EXPLORERS CLUB, YOU'LL...

√ Read fascinating first-hand accounts of adventure and exploration in our magazine *World Explorer;*

√ Have access to the **World Explorers Club**'s huge archive of history, archaeology, anthropology materials, and map room;

√ Be eligible to participate in our expeditions;

√ Receive discounts on **World Explorers Club** merchandise, travel related services and books you won't find anywhere else.

> If this is excitement to you, then you should be a member of the **World Explorers Club**, a new club founded by some old hands at exploring the remote, exotic, and often quite mysterious nether regions of planet Earth. We're dedicated to the exploration, discovery, understanding, and preservation of the mysteries of man and nature. We go where few have gone before and dare to challenge traditional academic dogma in our effort to learn the truth.

World Explorer Club • 403 Kemp Street • Kempton, Illinois 60946-0074 USA
24 Hr Credit Card Orders • Telephone: 815-253-9000 • Facsimile: 815-253-6300

Name:

Address:

Telephone:

Don't miss another exciting issue of *World Explorer*

[ ] **Regular Membership,** *US$25:* Subscription to the *World Explorer,* free classified ad, Expedition Discounts.
[ ] **Regular Membership: Canada: US$30:** Same as Regular, Airmail.
[ ] **Regular Membership: All Other Foreign: US$35:** Airmail.
[ ] **Couple Membership: US$35:** Same as Regular Membership.
[ ] **Contributing Membership: US$50:** All of the above plus a World Explorers Club T-Shirt.
[ ] **Supporting Membership: US$80:** All above plus a free book from the *Explorers Library* & a Thor Heyerdahl/Expedition Highlights video.

[ ] Check enclosed.
[ ] Please charge my VISA/Discover/Mastercard/AmEx.
Card #_____ Exp._____

One Adventure Place
P.O. Box 74
Kempton, Illinois 60946
United States of America
Tel.: 815-253-6390 • Fax: 815-253-6300
Email: auphq@frontiernet.net
http://www.adventuresunlimited.co.nz

ORDERING INSTRUCTIONS

✓ Remit by USD$ Check, Money Order or Credit Card
✓ Visa, Master Card, Discover & AmEx Accepted
✓ Prices May Change Without Notice
✓ 10% Discount for 3 or more Items

SHIPPING CHARGES

United States

✓ Postal Book Rate { $2.50 First Item / 50¢ Each Additional Item
✓ Priority Mail { $3.50 First Item / $2.00 Each Additional Item
✓ UPS { $5.00 First Item / $1.50 Each Additional Item

NOTE: UPS Delivery Available to Mainland USA Only

Canada

✓ Postal Book Rate { $3.00 First Item / $1.00 Each Additional Item
✓ Postal Air Mail { $5.00 First Item / $2.00 Each Additional Item
✓ Personal Checks or Bank Drafts MUST BE USD$ and Drawn on a US Bank
✓ Canadian Postal Money Orders OK
✓ Payment MUST BE USD$

All Other Countries

✓ Surface Delivery { $6.00 First Item / $2.00 Each Additional Item
✓ Postal Air Mail { $12.00 First Item / $8.00 Each Additional Item
✓ Payment MUST BE USD$
✓ Checks and Money Orders MUST BE USD$ and Drawn on a US Bank or branch.
✓ Add $5.00 for Air Mail Subscription to Future *Adventures Unlimited* Catalogs

SPECIAL NOTES

✓ RETAILERS: Standard Discounts Available
✓ BACKORDERS: We Backorder all Out-of-Stock Items Unless Otherwise Requested
✓ PRO FORMA INVOICES: Available on Request
✓ VIDEOS: NTSC Mode Only
✓ For PAL mode videos contact our other offices:

European Office:
Adventures Unlimited, PO Box 372,
Dronten, 8250 AJ, The Netherlands
South Pacific Office
Adventures Unlimited Pacifica
221 Symonds Street, Box 8199
Auckland, New Zealand

Please check: ☑
☐ This is my first order   ☐ I have ordered before   ☐ This is a new address

Name				
Address				
City				
State/Province		Postal Code		
Country				
Phone day		Evening		
Fax				
Item Code	**Item Description**	**Price**	**Qty**	**Total**

Please check: ☑
☐ Postal-Surface
☐ Postal-Air Mail (Priority in USA)
☐ UPS (Mainland USA only)
☐ Visa/MasterCard/Discover/Amex

Subtotal ➡
Less Discount-10% for 3 or more items ➡
Balance ➡
Illinois Residents 6.25% Sales Tax ➡
Previous Credit ➡
Shipping ➡
Total (check/MO in USD$ only) ➡

Card Number
Expiration Date

10% Discount When You Order 3 or More Items!

Comments & Suggestions	Share Our Catalog with a Friend